KATHREMES

KATHREMES

Universe
Magnetism and Levitation

Ulrika Skog

Homepage: ulrikaskog.net
Youtube channels: *Kathremes* (english) and *Din inre resa* (*Your inner journey*) (swedish)

Kathremes is a new revised version of *The Path to the New Earth* from 2022

Book Cover by Ulrika Skog
Translated by Mikael Dabrowski
Insert by Mia Fallby, m-Dsign.com

Publisher: BoD · Books on Demand, Stockholm, Sweden
ISBN: 978-91-8080-102-7

Print: Libri Plureos GmbH, Hamburg, Germany

FOREWORD

Nikodemus begins:

Why, then, are we writing a book about the sacred power of the universe? It is because the universe does not quite behave as you have been taught. Therefore, both now and in the future, we will be writing some books because parts of the information you have been taught have been distorted to teach you incorrectly. For this reason, in the future, we will use words instead of calculations to prevent others from creating more encrypted programs where only they have access to the correct information.

More scientists, both men and women, will emerge in the future to take over where, for example, Stephen Hawking ended his career. It is crucial as we enter a new era to correct the errors you have been given. This is our main purpose: to rewrite some books, to rewrite some knowledge that others have tried to obscure, which has happened so that you do not develop in a more free and independent direction. But do not worry, our power, the power of love, will always prevail.

In this book, however, we will only share some new information; the rest must wait until the time is right. Therefore, it is essential that we keep parts of the old information for now. So, use the information sources you have but try to take in the new, is our advice.

The book, therefore, only touches on the core of what, for example, an integrated coupling system is and consists of, because you must first deeply understand what time and space are before you fully understand what an integrated coupling system is at its core. But we will further develop our integrated coupling system and much more in the future, where we will clarify the parts that need to be explained. For it is from the sacred power of the universe that we will embrace the future and eventually learn to grow as a united people.

Furthermore, Ulrika has channeled all the information, both from her higher self, from us angels, and from the Kathremes. Therefore, this information is not something she should be held responsible for. She is writing a book, but the information is *our* responsibility. It may sound strange that an author should not be accountable for what is written since most have to be. But in this case, we believe it is necessary because much of the information we give her she has no knowledge of herself.

Thus, we do not dive into subjects that may contradict some questions and the answers that science needs today. We would not leave Ulrika in this sea of questions. Therefore, it is important and for everyone's best that we do not delve too deeply but only share what we receive from our highest Lord.

At the end of the book, there is a presentation that contains information about who the Kathremes are and how it is that we are writing together. Feel free to start with that part to get a better understanding of who they are and their knowledge.

The book was written during the years 2020 to 2021.

THANK YOU

Thank you, my beloved ones. All of you, archangels and angels who teach me everything about life. Your endless lessons are always available if we just dare to change, if we just dare to face our pitch-black inner selves. My guardian angel, so wonderful and pure. You have always stood by my side. And to the guides, thank you for your everlasting support and the knowledge you impart.

Thank you also to you, our dear friends the Kathremes, for helping us humans during these difficult times. Thank you as well to you, Stephen Hawking, my amazing teacher, for the strength you always give me and the teachings you are here to assist us humans with. It is true love that goes straight into my heart.

I love you more than life itself.

Thank you also to you, my beloved husband. Thank you, my dear one, my friend, my husband, my everything, my most important source of nourishment in life. A support when the winds pick up. When worry and weariness sometimes appear, you are there with your open, gentle arms. Arms I always want to live in. Arms I always feel safe in. Thank you, my dear, the man of my life, who made me blossom in the embrace of love again. I love you today. I love you forever.

Together we are strong, you and I.
You are my sunshine!

TABLE OF CONTENTS

INTRODUCTION

MY STAR ROOM

Ulrika speaks:

Here I stand now in my star room, in the middle of the universe, and within, I speak to my father (Nikodemus is my guardian angel. He is my ancient father, which is why I call him father). It is almost time for me to descend once again into a new earthly life, and the feelings I had then, I now convey to you, taken from my own reality, from the time before I descended to Earth.

I remain in my star room and I am filled with an inner feeling about the journey we are about to undertake, that I must try to reach my own reflection. To try to reach my inner self and illuminate the soul, the person who is me, my inner true self.

My father steps in and we converse:

– The time has come now, my child, for you to enter a new earthly life again. You will live with mental health challenges, but only so you can feel what it is like. But above all, you will turn inward for a while, which will give you time to heal and grow into the person you are meant to be in this earthly life. Then loneliness can be a beautiful and welcome feeling to reach.

You will also seek validation on Earth because it is a great era unfolding, where people increasingly seek outward rather than inward. For humanity is lost and suffering inside. They have been blinded by the external and the era of materialism that is taking place at that time, making both humans and Mother Earth very sick. We are going to help them with this, so they can find their way back to themselves again. That is why you need to experience what we will write about later.

– Thank you, father. I also understand that it is the only way for us to continue growing together so that we don't have to separate again, as we had to once when another guardian angel came into my life, which you can read about in our book "When the Angels Took Over."

– Yes, exactly. We shall now and forever live and work together, you and I, my child. For your soul has now come to where it should be and thrives in God's love.

– Oh, how wonderful, dear father. It warms my heart so much. I already long for you.

– And I for you, my child. But do not forget, we are always here and will always support you when you fall. For you will fall, sometimes very hard, but then we will lift you up again, so do not worry. Your husband will also be by your side, and you by his. This has been part of your plan for a long time.

I remain in my star room. There I stand in a vacuum of dazzling phenomena, as if the universe envelops my entire soul. I tune into Earth's frequency and hear earthly scientists and scholars conversing. They speak in riddles. It feels both strange and worrisome because they use words I have never heard before.

Father, they say the universe will one day go dark. But we know that can never happen, for then our soul would die. Then everything we have built together would collapse like Roman ruins, and the meaning of life would end. But father, why do humans speak like this? Why do they discuss a world they know so little about? How can we make them understand that everything we live in and strive for is eternal? That the universe will never go dark. How can we convey this knowledge and help them understand?

I look down toward Earth again. Father, I am cold, yet I am not afraid. I am cold from my inner sorrow, seeing how people keep watch over each other as they do down there. All the turmoil plagues me. It hurts so much when people hurt each other. How can we get them to stop? I cannot bear to watch!

I dread descending again. To once more let my soul endure the inner torments I need to experience to learn. But I also know it is the only way to reach a place closer to God. And I will do everything in my power, and more, to make it happen.

It is time now, father. We will talk again, you and I. I know this because you have said so, and that makes me feel safe. Farewell, father. I love you more than life itself …

… I have descended now, father. We have begun our journey. We have started our teachings. But there is no love here. The love I long to find, like in heaven, like in my star room. But now I have learned that not all love is genuine. Not all love is true, like the love between you and me, father. But now that we have found each other again, I can finally fill the void that needed to be filled, giving me the strength I need to continue our journey.

I have lived my childhood, and we have written our first book, the one that was finished long before I was born. But it is beginning to take a toll on my body. All the lessons have started to exact their price. They torment my inner soul, and its exterior has begun to show clear signs of ill health. But I will be strong, father. I will heal myself so that we can continue our journey, so that we can help and try to heal both humanity and Mother Earth, a mission our loving God has given us. And together with the Galactic's and the spirit world, we can achieve miraculous wonders. I want to believe that. But what do we do now, father?

– We will now continue our journey and venture into the universe, the world we all come from. For it is there that you will further develop your soul. It is here that your highest learning exists, my child, to assist Mother Earth and convey the knowledge you have gained in the universe. We begin with the moon and your soul.

PART 1

HOW WE CAN LEARN FROM THE SACRED POWER OF THE UNIVERSE

CHAPTER 1

THE MOON

INTRODUCTION

The Angels speak:

Before you can understand how Earth functions, you must first understand how your own energy flow works. That is why we start with the soul.

The energy waves of both the sun and the moon, and how they travel, will reflect how far we can travel in the future – when we use the power of the universe to transport ourselves. For this reason, we begin with how energies affect your soul. If you understand this, you will also understand the energy flow between Earth and the entire universe.

It is through the knowledge of yourself that we can help Mother Earth. And since we always work from a fundamental perspective, this is where we start, and in this case, we are talking about the moon and your soul. However, this will not be an in-depth study, we are providing only a simplified foundation to help you understand how everything in the universe is synchronized. After this, we will move on to the moon.

THE MOON AND YOUR SOUL

Archangel Haniel speaks:

The soul is always influenced by the moon, and it is through the moon's power field that you experience this influence. Many people will soon feel the increase in energy flow that the moon will release, both to demonstrate its power and clarity, and to assist Mother Earth with the strength she needs during her healing process.

The soul is also affected by the various moon cycles, whether it is a full moon, a half moon, and so on. The energy waves that come through during these times can be used to understand both your own and Mother Earth's energy fields.

– Some say they are not influenced by the moon. Is that true? I ask Haniel.

– There are many who claim that they are not affected by the moon. That is what they believe, at least. But everyone is influenced by the moon's gravitational pull because the soul is affected by it. The difference lies in how well you have learned to listen to your inner voice to connect with the moon's emitted energy.

It is not so surprising, really. We live in a stressful world where most things focus on the external. And if we don't have time to get to know our own inner selves, it becomes difficult to feel and understand others' inner selves, including the moon.

This is also why we have so many light workers on Earth today who are on their personal journeys. They try to understand their own and others' inner selves, and then they share their knowledge with others, for example, about the moon's integrated feelings and how we can harmonize them with the feelings of the sun and the Earth.

We refer to it as "feelings" because it is in your inner self, in your inner emotions, that the moon's gravitational pull meets your soul. Everything that a person feels always first lands in their inner self before they experience the effect externally.

– Why does the moon affect us, especially during a full moon?

– As with everything else when we talk about the universe, we always discuss interaction. The universe can never be governed by itself alone; everything happens in harmony with everything else. So, the fact that we can feel and be affected by the moon is rooted in the power field of the universe. Everything is an interconnected chain. So, you are not only feeling the moon.

The same applies to us in the spirit world. It is as if everything becomes day even though it is night, as we are also influenced by the wonderful gravitational pull of the full moon. We also try to take advantage of the moon's power by replenishing with new energy. In fact, what happens during a full moon is that both the soul and Mother Earth replenish with new energy.

If we didn't do this, we would go around feeling empty. Therefore, the soul cannot live on Earth if it cannot integrate with the universe. Your soul is created from the universe's sacred power, so without contact with something grand from the universe, the soul would perish. Thus, it needs to live on a planet that can integrate with the universe for its own survival.

– Don't all planets do this?

– That is true. Here, we only want to explain the significance, so that humans understand how crucial the moon is for the soul's survival.

– What about animals? Their souls are also created in the universe, right?

– Indeed, but for a soul to live and operate on Earth, it needs to follow the pattern that Mother Earth has established, which animals already do. It is also due to this pattern that animals will always surpass humans, despite all your machines and knowledge. Let us use the butterfly as an example.

BUTTERFLY

Archangel Ariel speaks:

Most people are already aware that butterflies can fly. But how many know that they can also act as satellites? Indeed, that is what they do. Besides enjoying the beautiful harvest of life, their role is to fly around and function as satellite receivers. Additionally, it is from the butterfly that you will learn how to capture the sounds of the universe in the future.

– How can a butterfly act as a satellite receiver? I ask Ariel.

– They do this through their sensory antennae. It is through these that they pick up the universe's signals. This is also how they navigate. Butterflies do not have *sharp vision*, as we call it. They cannot see where they are going next because they *feel* where they should go next. Therefore, they always know in advance where they are headed.

This type of self-guided navigation can also be learned by humans – how a butterfly's graceful nature allows it to move effortlessly without any conscious effort to steer itself. It simply follows its built-in navigation system. But it is primarily the universe's energy waves that enable the butterfly to navigate on its own. Butterflies, like many humans, are highly sensitive.

Have you ever noticed how some children are drawn to butterflies? This is often because the child is highly sensitive themselves, and it is their sensitivity that makes them, like some people, attuned to the moon.

Just as Ulrika experiences, who sometimes suffers from her heightened sensitivity. But the butterfly does not suffer from its sensitivity because it understands that its sensitivity is beneficial. It uses both its sensitivity and the moon's energies to navigate.

– Excuse me, but I don't understand. How does the butterfly use its sensitivity?

– As mentioned earlier, it is the butterfly's antennae that detect the universe's energy waves, and to do this, it needs an inherent sensitivity. This sensitivity is located in its body; it is the central navigation system. One might think that the wings are what provide the butterfly's sensitivity, but that is not the case.

Of course, the butterfly needs its wings to fly, but the wings are merely a means of movement. Just like a car needs wheels to move, but it is not the wheels that create movement; it is everything inside the car that generates the movement. So, in essence, the wheels are used to apply the movement created by the interior.

The butterfly has another ability that comes from the universe's powerful energy field, and that is its delicacy. Sensitivity is one thing, but delicacy is another, and in this case, delicacy means being light. It is crucial for the butterfly to be feather-light; otherwise, it could not fly despite its large wings. Without the universe's gravity, the butterfly would not be able to flap its wings even once. Thus, it can also influence its own weight, the gravity present in the universe, and which the butterfly perceives.

Perhaps some say: But isn't it the wing strokes that make the butterfly fly? Yes, we would answer, that is correct. But where does the power that enables the butterfly to fly come from? It draws its strength from the universe, which it then pulls into its body, creating movement, which in turn increases the frequency, making the butterfly's wings move.

You will understand more later. You will develop a better navigation system based on the butterfly's abilities, concepts, and design. And the only power you will need to navigate is the one the universe provides you.

Moreover, it is not as difficult as you might think. The answers are already present in nature – as long as humans do not alter nature. If you follow the existing cycles, you will also achieve more advanced results.

The choice is yours, dear ones. Our advice is always to follow what already works, and then develop from there. For when you understand how the soul's energy system works and how animals cooperate with the entire universe, you will also understand how the universe's gravitational pull works. And when you understand that, you can, together with the sun's power, use it as a renewable energy meter – or wave meter, if you prefer.

As long as you do not understand
the energy flow of Mother Earth
or your own,
it will be difficult to find a better,
renewable energy flow.

THE MOONS GRAVITATIONAL PULL

Archangel Haniel speaks:

When humans discuss renewable energy, the focus is usually on the sun, wind, and water. Yet, why has the moon not received the same attention in this renewable energy group? The moon also contributes energy waves. So why does humanity overlook one of our most significant and powerful partners out there? The reason is, as mentioned earlier, that you do not yet have a deeper understanding of either your own or the moon's energies. Thus, it is challenging to achieve a sustainable solution because all parts are not yet in place. And the most crucial part is the moon's gravitational pull.

The gravitational pull provides new energy and maintains the balance between the sun, the Earth, and the moon. It acts as a sort of *balance stabilizer*, both providing and receiving energy that can sometimes become too strong for humanity. Therefore, we need something that maintains this balance.

The moon also causes us to rotate, and it is this movement that we seek in renewable energy. It is the movement, the energy waves emitted by both the sun and the moon, that the moon's field can maintain, particularly when energy becomes unbalanced. By unbalanced, we mean when the sun's rays are either too weak or too strong, or otherwise difficult to collect, the moon can restore equilibrium.

Thus, when researching renewable energy and using the sun as a helper, you also need to enlist the moon's assistance. The moon is surrounded by an energy field, and the energy waves floating between the Earth and the moon can help you restore imbalances that may occur when experimenting with renewable energy.

The Missing Link = The Moon

How then do we link the moon with the sun and harness the gravitational source that the moon can balance? Most people know that the moon has an outer layer consisting mostly of lunar seas, lava plains, and highlands. But how many know what the moon's inner layer looks like and how it can gain additional energy from both the sun and the Earth?

Let's consider ions as an example: (An ion is an atom that does not have the same number of electrons in its outer shell as protons in its inner nucleus).

The moon consists of millions of ions, and if you could weigh them on a scale, you would see that they all weigh the same – there are no inconsistencies in what the moon is composed of. If you were to release the ions, they would land in a balanced way right in front of you if they could be seen and captured. This is what makes ions capable of acting as balance carriers.

– How can there be ions on the moon? I ask Haniel.

– Everything in the universe follows the same principle – giving and receiving. In this case, the exchange of ions happens through magma flow.

You see, for a planet to exert a strong gravitational pull, it needs to contain elements that facilitate this, which it does through the heat released by magma. This magma energy travels through various dimensions of the universe to manage and create a stronger flow, which is where the moon's gravitational pull comes in. It is through this that the Earth acts as a kind of energy source for the moon, which then converts this energy into other ions, which in turn are sent back to the Earth. Mother Earth then manages and distributes these ions in nature, ultimately benefiting humanity, as long as we live and work within nature. Therefore, there are ions on the moon, thanks to Earth's magma. This is how it works in the universe. Everything collaborates.

We do not mean that magma physically travels between the Earth and the moon; rather, the moon senses the Earth's powerful energy flow, which it then uses for its own purposes. This is how everything communicates in the universe, despite the distance. But distance is no disadvantage in the universe.

– Why magma specifically? And isn't lunar magma cold?

– We understand your question, and we will try to explain as briefly as possible. It is magma energy that we are concerned with, and as some already know, the energy does not need to be warm to generate power.

Once upon a time, there was indeed a beautiful flow of magma on the moon. However, over time, it transitioned to a colder climate, causing the magma to solidify. But just because it solidified does not mean that the energy dissipates; it remains functional. Otherwise, neither humanity nor the Earth would feel the moon's effects.

– Is there anything else on the moon that constitutes the basis of its gravitational pull? After all, isn't it so, dear ones, that the collaboration occurring in the universe also affects the moon's gravitational strength, whether it is weak or strong?

– Yes, indeed. Everything in the universe is governed together, and because everything integrates with each other, they also influence each other. What we mean is that if Earth did not consist of minerals that the moon needs to act as a gravitational source, it could not do so. Therefore, all planets in the universe are created with different characteristics. Otherwise, they could not benefit from each other.

Imagine if the entire universe consisted of a single, finished spice packet where everyone received the same contents. What would we learn from that? And how would we integrate and benefit from each other? Thus, it is crucial that the moon has ions; otherwise, the Earth could not benefit from the moon's gravitational pull as it does today. And if Earth did not contain magma, the moon would not experience the collaboration between our worlds.

As you can see, everything in the universe is intricately planned down to every little cell, every tiny millionth, as we call it. Otherwise, it would not work. And this is what we want to convey to humanity: that you not only use the sun but also consider how the sun integrates with the Earth, the moon, and the rest of the universe. And when you study the soul deeply, you will also understand the moon's energies deeply.

Moreover, the moon consists of much more than just ions. We only mention ions to highlight the messages we want to convey about magma energy and how you can learn how everything integrates with each other. Just look at how Jupiter acts and sends ions to its nearby moons.

Jupiter

Even Jupiter's moon Io emits ions from its volcanoes, influenced by Jupiter's magnetic field. Jupiter's magnetic field then transports these charged particles from Io to its neighboring moon, Europa, which is an icy moon. We call it an icy moon because a large part of it is covered in ice. This icy moon then undergoes its own changes, thanks to Jupiter's gravitational pull.

For many scientists, this icy moon would not be composed of anything other than ice because it is so far from the Sun. But it's not always the Sun that melts particles. Gravity can also act as a heat source, creating friction through gravitational forces that, in turn, heats up and melts Jupiter's moon Europa from within. Everything works together.

This is how all particles travel and interact with their surroundings – just as the Sun, the moon, and the Earth do.

Summary:

The moon's gravitational pull fluctuates with the flow of energy waves. Particles eventually land on Earth, such as in magma, which then transfers magma energy back to the moon through the energy fields of the universe, which is also what the moon needs in return.

Examples of substances include Nitrogen – Sodium Chloride – Phosphorus – Iron (magnetism) – Magnesium – Iron oxide. Salts and metals are of the utmost importance because they form a strong foundation for both Earth and the moon.

IONIZED ENERGY LEVEL

Archangel Michael speaks:

The *ionized energy level,* which we refer to but which you call ionization energy, is a future concept for how we can learn to harness the power of the universe. Therefore, ions, including those found on the moon, will play a significant role in our future fuel system as a valuable energy source.

Before we dive in, we first want to explain why ions could provide us with a better environment. To do this, we will let one of our distinguished speakers, our newly arrived spiritual guide, share his part. This is a role we have assigned to him because his knowledge of the universe is vast. However, we cannot reveal too much or provide overly advanced information. We only share what we are permitted, no more, no less.

God's spiritual guide says:

Thank you for your trust. I will, of course, do my best to share the knowledge I have gained and witnessed in the universe, at a level that humans can comprehend. But before I can explain how the flow of ionized energy levels will work in the future, I need to first address the power possessed by neutron stars and the structure of atoms. These are two separate points, but for now, I will only touch on ionized energy levels (we will discuss neutron stars and atoms later).

- We need neutron star technology to convert and create movement.

- Electrons are located on the outer shell of an atom, while protons and neutrons are inside.

So, how can we reconcile both? And how can we use this powerful phenomenon in the future? Furthermore, why are we talking about atoms, electrons, protons, neutrons, and neutron stars in this context? And what exactly is ionized energy level? To explain this and for simplicity, we will start with what an ion and energy are.

- An atom always has the same number of electrons as protons. However, it can also be transformed into an ion, which occurs when electrons either leave their shell (creating a positively charged ion) or gain additional electrons (creating a negatively charged ion). An ion is thus an early atom with an electrical charge and does not have the same number of electrons as protons.

- Energy is the part required and released when electrons are removed from an early atom.

Thus, the name ionized energy level.

When electrons are released, they are either affected positively or negatively, and it is from this event that a change occurs. However, whether this change is positive or negative depends on the state of these fundamental phenomena at that time.

We can call this the *transforming process*. The transforming process involves the release of electrons, which expands the circuit. When we expand the circuit, we can also integrate other elements that will be part of our future fuel system (we will discuss this further).

– Why do we do this, and what happens to the electrons when they are no longer present? I ask God's spiritual guide.

– They remain. They have simply moved to another space, where they now compete in a different domain, where they can merge their power with other parts. Thus, there is a transformation of electrons.

– Are there no electrons left then?

– New ones are constantly being formed, and this is also part of the transforming process. It becomes like a cycle where everything eventually goes around and around. New ones are formed – released, and then the whole process starts over again.

– Which is better, positive or negative?

– Both are equally good, because in this case, almost all change is beneficial. However, if you can learn to switch between both, it would be the best approach.

– Why is that? Don't they get a bit confused?

– Haha, you're funny. No, the world, with both positive and negative, has always been divided by humans. Why, we wonder? There are many good examples where both positive and negative can integrate with each other. If you let them work together instead, the power will be even greater than it already is. So, including both within the ionized sphere is a good path to follow. By doing so, you will also find more applications and interesting items (hint).

Therefore, it's important that we retain the electrons while adding other elements because if we transform the atom into something else or completely release it before adding other elements, the atom cannot act as we intend. No transformation will occur. Thus, to utilize this transformation, we need to study ionization at a higher level than we currently do and then implement the same technique in our vehicles.

How do we do that? Well, if we take an atom, for instance, one that is to act as a heating element, you will soon notice that the heat diminishes because the energy does not last forever. So, to keep the heating element positive and ensure the heat persists, we first need to transform the atom into what it was originally meant to be. Therefore, we transform an atom so it can act with the right energy for the right element; otherwise, it loses its function.

– Excuse me, but sometimes you want us to expand before we can merge. But here, we should retain the electrons before adding. What do you mean? I don't quite understand.

– We mean this: We merge *after* we have expanded and added other important components. It is only after a transformation has occurred that we can start merging all parts again, which increases our energy capacity. Do you understand, Ulrika?

– Thank you, now I understand.

Summary:

Electrons are released, but only a bit outward, to expand the circuit. This means they are still present but in a different space because they have moved. After this, we can add other important elements. This is where a transformation occurs, where one substance can be converted into another. Then we merge all the ingredients again. They become stronger, which increases the capacity.

Here, you also need to get better at using magnesium in your container, because without magnesium, you cannot start your journey. Therefore, the technique needs to proceed step by step, as without magnesium or other essential components, you will not reach the future's ionized energy level. And remember: everything collaborates. Even if some positives do not always combine with negatives, try anyway! Continue to act as a positive channel in a negative world, and you will also be able to link your ions together.

Nucleotides also form a strong foundation in this context, and together with ions and magnesium, you will soon understand what we mean. Nucleotides will play a significant role in the cell's energy storage and transport, learning to both remember, store, and transport. They will act as a memory bank, so that the energy knows how to store and manage its contents the next time.

– How can nucleotides form a strong foundation in ionized energy levels?

– You see, if done correctly, we can indeed store an ionized energy level, but not in the same way we store energy in a battery. Here, we advocate for storage that occurs when we supply energy to achieve ionization. This is something we can apply to our nucleotides as well. But it's not the building blocks we are after. We refer to the technique used as a means to store, remember, manage, and transport energy.

IONIZATION ENERGY WITH INFRARED LIGHT

Kazandra speaks:

To store and transport energy, we also need to reach an ionization perspective. Specifically, we can multiply an ion – split it into two – while using infrared light. This process achieves an *ionization perspective.* This ionization perspective can then be used to convert new current in a circuit so that the power can continue to flow.

However, to accomplish this, we need an accelerator that can channel infrared light. Infrared light provides additional strength to the ionized unit. Thus, the accelerator acts as a gear shift, adjusting the power, for example, in a car.

For ionization to bind together, it requires something that can add its own power, and this is where the ions come into play. As power increases, ions also increase their internal strength. They then transform into free-moving particles that travel along a specific pathway.

To start and maintain the accelerator, we must continuously double the power, which leads to ionization energy in infrared light. We then have ionization energy – a new force that can multiply itself through various current-carrying circuits. These circuits, or rather pathways, can then be used to stimulate the power. This is also where your infrared light comes into play. By using light, you can bind the multiplied power and add additional power to your new project.

We also need a transistor. It will be the component that transfers power from the outside. Hence, a transistor is needed to reach external forces and transmit the power. The transistor is placed outside the accelerator and is responsible for conducting the power. As the particles increase in strength, the transistor will detect this and transfer the power to the accelerator. This results in a power increase, meaning further transformation occurs, which then multiplies itself.

The transistor functions as a guide/mediator and is crucial to include in our future project. This will not only create an entirely new light but will also significantly enhance the power within the ionized energy level. Therefore, we need a transistor – a transmitter.

This power is self-regulating but requires assistance to multiply itself, which it does through an infrared heating device. You will create this device yourselves, and when you do, you will discover a completely new ionization force through this instrument.

This will be your next project: to activate a WARP of infrared light that can analyze new data and independently add its own power, which increases transmission – the power conveyed.

Here we are not permitted to go further. The rest is for humans to discover and achieve an improved version of the ionized energy level. You will also illuminate the moon's internal power in a completely new way.

THE MOONS INNER POWER

Kazandra speaks:

When studying the moon in light of today's issues or inventions, you use contemporary instruments. However, in the future, as you gain a better understanding of the universe's power, such as the sun's power, you will also transform the moon's energy reserves into a current-carrying link.

– How do we do that? Isn't it more challenging to integrate lunar power into an electrical grid? I ask Kazandra.

– Indeed, it is. The reason many might be scratching their heads now is that you haven't yet created the instruments needed to achieve this. That's why some are hesitant. But trust us, it will come. The moon will take over when you can no longer access the sun's power in the same way you do today, which we will discuss later. At that point, the moon's time and understanding will increasingly come to the fore. However, only a highly spiritual group of souls will understand and develop the concept of *using the moon's power as a current-carrying link.*

– Aren't we already capturing the moon's power? And why spiritual beings?

– Not in the way you will in the future, because you lack an inner understanding of how the moon's power can be distributed differently. To access this inner power and manage and use it on Earth, we need souls who can delve into the moon's internal energy.

For those drawn to the moon both spiritually and scientifically and interested in its significance, we encourage you to delve into this learning and further develop the moon's inner power so that it can also serve as a future current-carrying link.

– What is this inner power? Can you tell us more about it?

– We can say this: First, you need to understand the moon's journey, which he himself will explain later – how he emerged and established a good connection with Earth. It will be this connection, this interaction, that these individuals will study. Once you do this, you will realize that the moon's inner power is much stronger than what is mentioned in your textbooks. That's why the moon's power is not used in the same way you use solar power.

It's "they" who have obscured the moon's true power, only highlighting its exterior. It's a power source that can be harnessed similarly to how you use volcanoes on Earth.

This is the power we are referring to, and it is very strong. Understanding it requires a high spiritual elite because it involves considerable knowledge to handle it. Thus, we need new instruments. But it will come, dear ones. Everything will eventually be revealed, and then you will understand what we mean.

– For what purpose will we use the moon's inner power?

– You will produce according to the moon's way of operating, where power is emitted from within. This is the discovery you will make and link with a current-carrying concept.

– Why the moon?

– The moon is a wondrous place with more to reveal than you realize. But the knowledge you possess today is insufficient. The moon is much more compact. By compact, we mean it is filled with valuable insights you can learn from, and this will come when you better understand the interplay between Earth, the sun, and the moon.

– So, can we calculate the moon's power scope, just like the sun's?

– Not entirely. When it comes to the moon, you will not calculate the scope; in this case, it's irrelevant. The calculation, or shall we say execution, you will undertake in the future goes deeper than that. But don't worry. Some countries will start collaborating on this, and we also hope that you do so and share all you see and learn.

Archangel Michael speaks:

When working with information regarding the universe, it's crucial to share everything equally, which doesn't always happen, especially with space travel. Therefore, our friends, the Kathremes, will assist us in this endeavor. They will oversee the souls who wish to retrieve information directly from the universe.

The reason for this is that some people withhold information, including those seeking *early* information to use as signals on Earth in a highly negative way for us. Therefore, we have now put a stop to this. As long as it is not done in a peaceful manner, we will be there to influence!

There are also those who do not share their discoveries. They claim to, but they do not. This too will come to light in due time. In the era we are currently in, much new information will be leaked. Those who claim to always tell the whole truth will be revealed. There is a considerable amount of information about the moon that some researchers possess but do not share.

Everything that has been said
has been written down,
but everything that has been written down
has not yet been spoken!

Therefore, look into their desks. Light-colored, old, and worn in appearance.

That's all we can reveal, but we want you to know that a more truthful elite will take over in the future. They will review their old colleagues' work and then post what they have found on the internet. Much of what has been said comes from inaccurate sources. Thus, we now let the moon step in and reveal who he truly is.

THE MOON SPEAKS

Archangel Haniel speaks:

The moon now tells of the collaboration that will occur when Mother Earth begins to heal. This is a prime example of how everyone in the universe always helps each other. Therefore, we will now take you into our world and tell you about the moon. Thus, we have now awakened the moon's energy distribution in Ulrika's soul, so they can meet and assist humanity with what you will need help with in the future.

Ulrika speaks:

Through an astral journey, I entered and reached the inside of the moon. Below, I write down the event, and the moon speaks in its own words, without my correction of the text.

I am drawn towards the moon. I float in and pass through the outer material that makes up the moon. I penetrate and reach the center, the inner core. It is small but powerful. Much stronger than humanity understands. It also rotates much faster than humanity comprehends. All movement, all vibrations that the moon emits come from the inner core, and that is where I am now, deep within its recesses.

I look around. It is quite still, despite the powerful energy present here.

I make my way out again and float up towards the universe. The moon shows me a completely different picture than what we see on TV. He shows me rings. I don't know why, but once upon a time, there were rings around the moon, small but very active.

– What happened to them? I ask the moon.

– They disappeared when the gravity in our solar system changed direction, if we may call it that. They were here to create and bring out my inner self, but as the universe changed, so did I.

– How did that happen, and when did it occur?

– It was a major change that took place before Earth was set in its orbit. That's where we then met, Earth and I. And by losing my rings, I could sail away and meet Earth without obstacles, which was beneficial in that way.

Let me explain:

Previously, when I had rings around my body, the path was

actually different. Back then, it was said that I would stay further out. My rings were meant to increase the power I had inside and create other adjacent elements around my body. But when I entered your orbit, my routines changed, and I instead met other bodies that assisted me then. When I lost my rings, it opened up for a better meeting between our worlds, and I could reach Earth more effectively. I met Earth, as you know, and we started to collaborate instead.

So, it was an event that occurred and was originally meant to be something completely different. But the result was that I could help humanity in the work we have ahead of us now, in the new age. When that comes, my power will change, enabling you to open up more than you have previously. This was what God had planned back then – that the current era would arrive one day. Thus, he laid the foundation for humanity's journey even then.

This is my journey, and as mentioned, I will help you in the future under a different concept. Although this is ahead of its time, learn to embrace my inner power as you have never done before. Then you will better understand the new Earth's inner concept, for that is where my power lies.

– It sounds like it ended up being a good solution for both of you.

– Yes, everything our Lord created for us, and as my character developed, our collaboration could begin. We started to collaborate because of the gravity we mentioned earlier. It was among other things that brought us closer. I am essentially a mass that flew here to help you in your solar system, but where I would land was not certain in my realm.

– Wasn't the moon created from pieces of Earth? It is said that the moon and Earth collided with something, and then the moon emerged.

– No. This is misleading information. It is true that a collision occurred, but it did not affect me. Admittedly, I am made of pieces of Earth, but it's only my exterior. My interior was already available to the entire universe before my rings were created.

It was when gravity increased that I could finally transform my power, and my interior was propelled into the absolute orbit where I now reside. There, I could manage, rest, and establish a collaboration with Earth. And as my power increased, Earth's magnetic force also increased.

– How did Earth acquire magnetic power?

– Earth's original power is created from a power source; otherwise, no one would survive in the universe. For no power, no flow, and so on.

– When did your collaboration expand?

– It expanded as more came to us within our solar system. The more we became, the stronger our solar system became.

– Do you mean that Earth and you were there before others?

– No. I do not mean that, some came before. What I mean is always the power that gathers when we become more. Before that, the power was weak, but as mentioned, the more we became, the more powerful our solar system became.

– What will happen now and in the future?

– Now I will go in with my inner power source and transmit it to Earth. This is to assist with the gravity that will be released when the sun's powerful rays increase. When these rays increase, the risk also increases that everything gathers at a single point, which is not good when we talk about gravity.

I want to clarify my response:

The fact that my inner core isn't as powerful as many others' is mostly because I sometimes need to release my energy. However, for several years now, I have been conserving some energy levels to be able to meet the significant changes that will happen on Earth now. Simply put, I will be helping you with more power than I usually do, and many of you, my friends, will feel this. Therefore, my advice is always to connect and speak with Mother Earth, spend time in her nature, drink and consume fresh energy. This will increase your ability to handle my new energy.

– How long will this last?

– Within a year, I will release my energy again. Then my part will be done. After that, other planets will step in and help you during the second year. When they come, they will bring you fresher breezes. This is to cool down your instruments again; otherwise, some manufacturers will overheat during this period.

Among others, we will call in Pluto. Pluto will show himself more than he usually does, and what Pluto will assist with during that time is integration. By integration, I mean bringing everything together again, and that is the task we have assigned to Pluto. Humans see Pluto as an unworthy member, which we do not, and after this, you will put Pluto back on the map.

After that, other planets will take over and cool down Mother Earth.

The Moon continues in his own words – who he is and what he does:

It's also time for me to convey my extensive knowledge to you humans. For humanity has long wondered who I am and what functions I can bring to this marvelous solar system. I will share some of that now.

For me to function, I need to rotate. My rotation is not really visible, not entirely at least, but I do it slowly and surely. I bring about rotation because it creates a free flow in the realm where I reside. It covers the parts I don't share. But it also conveys the knowledge I get to impart. So it's a dual rotation (force). That's why I rotate. But it's not something humans can see with the naked eye. It's an inner rotation, an inner force, that affects my position. Everything I do always come from the inside.

Therefore, it is now time for me to step forward and clarify to humanity my true nature and the function that will bring new light into your daily lives. For your light will soon change its character, and it's within my inner self that you need to seek (a clue). For it is always within my inner self that I exist and convey all my inner wisdom.

It is a wisdom that has attracted many visitors to me. Some have succeeded, while others have been less successful. But those who have understood who I am have only reached past knowledge. Therefore, in the future, I will loosen some new reins, and then some of you will get to experience the future's merlot. The future's merlot will be presented by more who want to connect with me, and thus more will learn from my power.

By merlot, I mean merging objects so they eventually leak out. In this way, we connect with the properties of your grape merlot and the process that occurs when you press out all the content. The same thing here.

But it is not yet time to convey. In this matter, I only step in to warn those who are here to expose their guilt. For I will no longer accept those who rule with means other than the good. I will no longer allow that to happen.

So if you do not serve the good, I have been granted permission by our creator to extinguish your power, and many will help me in this. So if your mission in my realm is not of a good nature, you will no longer bring news home to your family. But if you work for the good, I will gladly assist you and yours in your work.

This is what I want to convey today, Ulrika. Sleep well, my beloved child. For sleep is important, and that you sleep all night long. Thank you from us, the Moon, and all of us who surround you with a good atmosphere in our entire solar system. Now my friend the Sun will take over.

CHAPTER 2

THE SUN

THE SUN'S GRAVITATIONAL PULL

The angels speak:

What kind of power exists in the sun, and how can we learn from it? What do we really know about this massive star that daily spews out fireballs to avoid its own demise? And how reliable is the sun, really?

Additionally, how can we harness the heat and gather all the power the sun emits so that we can store and use it later? We already do this through solar panels and in a type of generator where energy can be stored like in a pantry. But how sustainable is that energy?

And why don't you combine the energy waves of the sun and the moon? The universe does. Why does humanity separate what the universe integrates? Why not link everything together instead? Especially now that you have learned that everything the universe links together increases capacity and the ability to act. But how do we integrate the moon's energy with the sun's? Can we use the moon's rays in the existing solar panel? Would that even be possible? Yes, we say, because it already happens like clockwork throughout the universe.

This is one of the things we want to highlight in this chapter regarding the sun's power.

It is the sun's energy, the energy that the sun's rays both manage and use for itself, that humans have studied, and which you already use as a power source. But to move forward, we need even more powerful energies than what the sun gives us today.

– Aren't the sun's rays powerful enough? I ask the angels.

– No, they are not enough. Because even though humans have finally found a new way to capture the sun through your solar cells, even beyond the clouds, it is not nearly enough because the sun itself is not a reliable source. It cannot be controlled, as the sun's rays can sometimes seem both *hard* and *shifting*, as we call it. Therefore, it is unreliable.

What we mean is that we cannot control the sun's hardness. It might seem like the sun is free to act as it wants, but it is not. The sun actually has a certain *obedience* to the universe, as we call it, and this obedience is constituted by its inner power. This obedience is essential as it dictates how much or how little power the sun releases. If it happens too quickly or if the sun releases too much energy, Earth would not exist.

Other governing phenomena do not have this obedience; they can float around freely and release their energy even in other places, even outside their own zone, as they are not locked like the sun is. The sun is locked to function as a link between us and other phenomena within our solar system.

By shifting, we mean what happens when the sun is influenced by other phenomena in the universe. When that happens, a shift occurs, and when it does, you cannot control how the sun's energy waves land in your solar cells. That is why you need the moon because the moon can balance this out, for example, via ionized energy levels (ionization). This is what we mentioned earlier.

We won't delve into this now. Here, we just want to explain why some of the sun's rays are not always entirely reliable. But keep shining, keep using the sun as a power source, even though other components will outshine the sun as a kinetic energy source in the future. The sun will always remain in our lives, even as a driving force, but on a smaller scale.

We continue with the collaboration of the sun and the moon:

What we advocate instead is a collaboration where the sun's rays emit heat and where the moon's energy is used as a driving component, making the energy last longer. And if it only returns to Mother Earth, and humans do not misuse its power and strength for their own sake, the energy will last forever.

It is the connection between Earth – the sun – the moon that humans need to understand better, and it is knowledge that is already on your desk.

But the most important knowledge you need to highlight now is: *How is it that an energy wave of this kind can reach so far?* For this will become the source of future energy fuel because it is what allows you to fly long distances. The energy waves are much stronger and more far-reaching than those we have on Earth. Therefore, it is essential to find the right balance between the energy waves – to ride the wave that comes and learn what is on Earth.

Moreover, humanity has already tried to fly on this energy wave, and it is precisely this that you will further develop in the future. The difference between now and then, as well as other future scientific developments, is that the energy we use today is not quite used in the right way according to us. But you are on the right track, so good job!

Energy waves = movement. Despite long distances,
we can learn to use their character.

How do we then access these energy waves that can stretch so far and for as long as the sun does? And what are the sun's energy waves really made of? What is the moon's energy field made of? But most importantly, what do they ride on, and what do they contribute to make this happen? Something for you to think about. But we will help you along the way, and to do that, we first need to go into the tilt of the Earth.

THE EARTH'S TILT

The angels speak:

Before we move forward and develop the universe's energy waves, we first need to understand why and how the Earth tilts and how this tilt actually enables the sun's rays and the moon's power to collaborate. But not only that, we also need to understand why the Earth rotates.

There are several reasons why the Earth tilts, and one of them is that Mother Earth herself has chosen to use it as her own function. Among other things, she uses the tilt to release energy that no longer serves her. There is an extra path, an extra solution, to solve Mother Earth's problems so that she is not affected by too much energy from the moon and the sun's energy waves, and that is by emptying out her excess. But this can only occur when the Earth is in a lying position, or as you say: standing. And it can only happen through the South Pole and when the Earth is in the right position – when the polar circle has shifted.

Everything in the universe can be captured,
but you need to be in the right place, in the right
position, just like the Earth is.

The moon also influences the Earth's weather conditions, which we want you to delve into now, especially when discussing future mechanical parts. Because many significant changes will occur in the future where the Earth's tilt will also affect our climate. Where it is winter now, drought will arise. Where it is drought now, there will be more rain. This is foretold and written in our future books, the ones we have in the spirit world.

The moon also stabilizes the Earth's rotation, as it is necessary for the Earth's development. If the Earth stood completely still, it would not only negatively impact the Earth, it would also negatively affect our entire solar system, and we would then have no energy flow. That is why the Earth rotates, and the energy required to get everything in motion can be used for energy consumption on Earth.

The moon keeps the Earth
in place – the Earth keeps
the moon in place.

How can we then use this method of action in the future? Because we will. We will adopt both the Earth's position and its tilt as an internal concept. This is what we need to improve and understand before we can comprehend future technology. If we are ever to harness the universe's power in the future, we need to first understand where and when it emits the most power. The positions of the South and North Poles on Earth can have entirely different positions in the universe, as the universe has a completely different determining effect.

So when we talk about the sun, the moon, and the Earth, and the power you need to harness to illuminate the Earth, you also need to consider the Earth's tilt. If you do not, you will not understand the flow that occurs between the sun, moon, and Earth, and how you can use both its gravity and energy flow in your daily life. To understand this, you first need to further develop your solar pattern, which Bermuda will now explain.

SOLAR PATTERNS

Bermuda speaks:

Hello, my friends. It has been a while since I last connected with Ulrika, as my friends and I have been on an exploratory journey. And it brings me joy that we are in contact again.

– Oh, my dear friend. I have missed you so much!

– And I you, my friend. Thank you for your love. We have all missed you dearly.

– Where have you been, if I may ask?

– My colleagues and I have been on an exploratory journey, and that is what we want to clarify now. Therefore, we will explain more about what is happening with the sun and how you can harness its rays in a completely new way.

The sun doesn't always behave as humans think it does. During the journey we recently undertook, we made not only a massive new discovery (which we cannot disclose here), but we also learned how to pattern the sun's rays, which you must understand and create first before you can develop a new energy flow (energy waves).

By patterning, we mean mapping out the path of the energies, which involves the pattern, the path, that the sun's rays take to reach the rest of the universe. You might not be able to see it in your telescopes, but the sun's rays do not always act the same way because they are irregular, hence unreliable. Their irregularity is due to the internal force the sun releases. Sometimes the internal force is stronger, sometimes weaker.

The sun's rays do not always take the same path, and this is what you need to learn first, even within the solar magnetic world. By doing this, you can also capture the sun's rays that you are currently missing, and you miss them because, as mentioned, they are irregular.

– How do we do that? I don't quite understand.

– You do this through a patterned solar system, one that can also capture what we call *inconveniences*, but for us, they are not. These are black particles that we sometimes find in a single energy wave, and they can be transformed so that you can also use them in the sun's reach.

We call them black only because you cannot see them, but we can. It has nothing to do with color, but it's a bonus if you can incorporate them into this patterned network of energy waves.

These particles will pose a significant obstacle in the future when you aim for pure solar energy. Therefore, you need to learn this and overcome this obstacle. When particles get caught in the solar pattern, they lose their power.

Think of a chimney. You apply a force from the fire (the sun), then when the ash (black particles) comes out, you can either filter them out or transform them. That way, you also get cleaner solar energy.

– Why do we need to remove or transform them?

– Because they slow down the speed.

– How do we do that?

– By understanding the solar pattern. Sometimes they are numerous, but it's also then that it becomes easier to achieve pure solar energy. You draw the particles in through a magnetic process.

– How do we transform the particles?

– Your professors will calculate that. But it is a transformation that takes time. Therefore, it's important to tackle one pattern at a time and try to understand what they consist of. We call them *nifoglyms*, but you can rename them as you see fit.

There is a pattern that the sun's energy waves always follow, even if they seem irregular. When they are then released from the sun, they are stationed at different places in the universe. Some energy waves reach Earth while others do not. Some dissipate immediately after release. Some are strong while others are weak. Even though the weak energy bursts do not reach Earth, you can still capture them by understanding the sun's eruptions. If you always know when even the weak bursts occur, you can adjust your own frequencies to meet them before they dissipate.

– Isn't that very far away?

– It is, but you will learn this in the future, how you can meet energy waves that have just been released so that you can use them as solar power. They are much stronger. Not in the long run, but at that very moment, at the time of release.

– How do we do that?

– You can't do it now, but it will come. Your scientists will then participate in what is called a distance marathon. A distance marathon is about capturing what happens at a distance, over a long period. This is what you need to do to capture more energy waves, even those that are far away.

You will also develop a new measurement instrument. This instrument will be capable of detecting and assessing when the next solar flare is about to occur. In this case, we are referring to short-distance flares, as you can already predict the longer ones. This instrument can *bend* in time and space, *pass* through obstacles, and *capture* energy. It can bend, pass through, and capture. Through this, you can harness the sun's inner power even more, including its magnetic capabilities.

We can't reveal more at this moment, but it will come. So, don't worry. Everything will soon fall into place regarding the solar power you need to improve.

And remember: When creating a solar pattern, design it based on your constructions, not the sun's. You need to work at the level where you are, not where others or the universe are. This solar pattern is so well-constructed that it will take generations before you fully understand its concepts. That's why it's crucial to start creating a solar pattern based on where you are right now. Then, expand the sun's pattern as you develop your knowledge. It's a solar pattern with many new pathways that you can use if you start mapping out this pattern.

– How does the solar pattern look?

– It's like a puzzle, where each piece, each step, each station, enhances your knowledge. Think of it as a chessboard, where each square represents a specific quotient of information and a particular epoch in your history of knowledge. When you understand what each square consists of, you also understand the foundation of the universe. Everything is meticulously drawn out, and each detail has its own unique knowledge and characteristics.

So, study this surrounding pattern until you reach the sun's core. Then, you can apply the same technique to your existing solar cells. That's why your solar cells are already divided into grids, where each cell represents its own history. But before you can calculate the power each cell needs and can absorb, you must first understand the full extent of the sun's power.

THE SUN'S EXTENT

Kazandra speaks:

It's the sun that will determine your future, both how you learn to listen to the solar flares and how you harness the right power that emerges at the right moment. When the sun's rays fade and burn out, more power remains than you currently know, and that is our next topic.

The extent of the sun's power is measured in amperes, if we may call it that. But it's an ampere that won't function in the future. By ampere, we mean the sun's strength, the sun's energy power. We don't mean a lightbulb in amperes; we mean the strength in amperes, the one recognized by the sun.

The reason you can't use amperes in the future is that the sun's power will become even more irregular, especially as its power wanes and the sun is replaced, which we will discuss later. Instead, you will use a regulator. This device informs you about both the sun's power and how close you can manage this power in your own projects. To calculate how much power each solar cell needs and can absorb, you first need to understand how the sun's power works in its entirety. Once you do, you can also utilize the power the sun emits before it dies out. That's what Bermuda referred to.

But to understand this, we first need to grasp the full extent of the sun's power, especially its inner power. The rays emitted by the sun are all created internally. The sun has no external power since all its power is generated internally. That's the part where the sun's rays shoot out their power like Cupid's arrows, and we can learn to capture this power early on.

How do we access this internal power? How can we reach power that is so far away? We need new measurement instruments. We need a powerful sensor capable of detecting the most elusive rays.

As Bermuda previously mentioned, you will learn how to capture even the sun's shortest rays, those that lose power much earlier than others. Since this power is so far away, we can't calculate it the same way as the long rays, as they fade much faster and need to be captured under different circumstances. This same meter can sense when they are coming, allowing you to capture even these in your future solar cell, in amperes. You can then input the entire range of amperes into your data, informing you when the next batch of short rays is due.

Think of it like your temperature gauge. You have an upper part representing positive, and a lower part representing negative. When the meter senses the sun's rays, it moves up to positive. When the short rays are far away, the meter is at negative.

We can learn to see and feel when and which sun rays emerge and at what stage, which can also be calculated, but not until you understand how the sun's interior works. Further research is needed for this, and once done, you will adopt a continuous calculation process that senses which and how strong the rays are before they emerge. You can then use the short rays, which are incredibly powerful in their early stages, in your solar cells.

This will be a
special instrument used to collect
the rays closest to the sun.

This is an internal knowledge that needs to be improved to understand the sun's external power. Therefore, our wish is for you to acquire more knowledge about the sun so you can further develop the technology you already use in your solar cells. To fully grasp the sun's internal capacity, we must first understand the power that the entire universe consists of. If we don't understand how the universe functions, we can never understand the sun's interior, and since the universe is not defined by limitations, neither is the sun's interior.

What do we learn from this? Everything we need to know about the sun and its interior already exists in the universe because everything there integrates with each other. To understand how the sun functions internally, you need only to observe how the universe functions.

When you see the connection between the sun's inner and outer workings and how they interact with each other, you will also understand the solar pattern we discussed earlier. This interconnected chain unites the sun's interior with its exterior, which is what we want you to delve into now.

As your pattern begins to emerge more clearly, another force will become apparent. It is a magnetic force already used in other solar systems, those that have advanced much further than you. But listen, this is an extra power source that you must calculate, as it cannot be observed. So, develop your newly acquired calculation system, for that's when you will discover a new solar energy and will thus double your power within a few years. The Solar Mathematical Association will discuss this later.

Following this, you will further develop your solar cells, and when you do, you will need new storage plates capable of managing the enormous power, the current, you have discovered. The old plates will no longer suffice, as using them increases the risk of them either breaking or exploding, which happens when the current cannot be handled by the material.

Once you have harnessed this new power, you will also develop the same technology for a transformer, which we will discuss in part three. Then, the solar plates will decrease in size to be usable within the minimal concept.

NEW SUBJECT

The angels speak:

To further harness the power of the sun, you will also need a new element. We will use folate as an example of the interplay between the sun and the Earth. It is a very old example of how important the transformation of matter is for us to achieve new, improved powers. Therefore, we will now delve into the composition that took place even before The Big Bang. For it is to that time we need to go to illuminate the structure of the sun and the moon and the fusion that occurred at that time.

Long before our own solar system was created, there were already numerous forms of matter in the universe. This matter served as a kind of foundation, a prerequisite or preparation for something else to be created or to happen, and one of these was folate.

– Why folate specifically? I ask the angels.

– Folate has many functions, including being important for our well-being, especially for women who are trying to become or are already pregnant. Therefore, folate has always represented rebirth and is also the reason why folate was created. Some might now sigh and argue that folate does not exist in the universe. But they are wrong, we say. It does exist; it's just a tougher nut to crack.

Without folate, we wouldn't have reproduction from the very beginning.

– What does folate have to do with the sun and the moon?

– Folate is just an example of what existed long before our own solar system was created. Without the matter that existed at that time, neither humans nor animals could live on Earth. When the big bang occurred, all the basic nutrients were already in place. After that, the universe could further develop its elements, even into other dimensions. But the foundation was laid, and the sun, moon, and Earth could be created.

So you see, everything that happens in the universe has never come about by chance. Everything is integrated, both in the atmosphere we had then and the one we could integrate with later. That's why you find the same elements on other planets.

– How was folate created, and how could it survive in the universe so that it could then be applied to Earth?

– Folate wasn't folate from the beginning. It was a mass in the universe that contained a substance that would become folate. After that, folate changed its character according to the needs on Earth.

– What does this have to do with the sun's powerful rays?

– After Earth was touched by various materials it needed, the sun took over. The sun's rays enabled the elements that existed on Earth at that time to shed their previous shells and renew themselves to become, for example, folate, which is crucial for life to arise. This is something you can also do in the process with your solar cells.

– So if I understand you correctly, we can add an element to our solar cells where, with the help of the sun, the element transforms and increases the power?

– No, not entirely. You need to first find a new element that can be transformed with the help of the sun. Then you use it in your solar cells to further increase the efficiency because that element is better at capturing the sun's rays. It will be an additional underlying plate that you will place there, containing this element. This maximizes the use of your solar panels. So the sun is used in several processes before you are completely done.

– So we can transform an element depending on our needs, like folate?

– Yes, exactly, and that is what you will do, because that is what will increase the power.

– I get the feeling that the element is already here and is also intended for this very purpose, as a help to capture the sun's rays.

– That's correct. But remember, the element exists in its original form. For example: if you had looked for folate, you would not have found folate. So look for what *extends its power*, which can then change its power to increase its strength.

After that, you will develop a *finished* technology that can also capture the moon's energy consumption (hint).

So you see, everything that circulates eventually takes hold. We start with a matter that then changes its character, depending on its task. And that is how you will further develop your future technology. For if you learn to transform the energies of the sun and the moon into other factors, the energy will increase. Therefore, the transformation of elements is necessary and is also the purpose of our story.

Folate was thus a prerequisite for life to arise. Even though it initially existed in another form, it then gained a new identity on Earth. But that's the case with all new things. Therefore, it is essential, especially when discussing the universe, that we always learn from the ground up.

When you have found more of the parts that the universe consists of, at least those you need, you can also, as we mentioned earlier, use the sun's and the moon's energy levels in a completely different way than you already do. Then you will also understand what the sun is trying to send out, and then you will also know how to capture the right elements, at the right time, and with the right material.

Now, folate was just an example; of course, there is much more than that. Here we only gave a small hint of what exists and what more is out there if you just search a little deeper and a little further back in time. If you do that, you will also understand the content and connections of the sun, the moon, and the Earth, which you can then use to create a new energy flow, a transformative process, just like folate when it made its journey.

THE SUN WILL BE REPLACED!

The angels speak:

The power of the sun will thus bring forth new discoveries in the universe. What you couldn't see before, you will now see, such as your new element and why you previously couldn't measure the full extent of the sun. The truth will come to light here as well, and you will then sail toward a new horizon and finally unify your knowledge with each other.

Furthermore, we want humanity to stop using terms you don't fully understand yet, such as that the universe will one day burn out or that the sun will explode and destroy our solar system. This is merely scare propaganda! Therefore, we will now explain what we mean by the sun's power changing.

The sun has not always had the power it has today. Long ago, it needed its own world to develop in. This is the core of our sun's creation – that it once lived its own life before it could anchor itself in our solar system. All stars go through this process. Either they explode to release all their contents, or they grow larger to assist other planets within and outside their own solar system.

Believe it or not, there are many stars that exist outside their own sphere, where they form and process in what is called a pre-stage before they can act as a sun for a planet or an entire solar system. This process is necessary because, without it, we couldn't live on a planet as everything created needs an energy source.

– Are there sun's that act solely for a planet, or do they need an entire solar system? I ask the angels.

– A sun can sometimes give off its energy to help other planets. But ultimately, it usually ends up in its own solar system, where it is meant to be.

– How did our sun come into existence? And is it true that it will explode in a certain number of years?

– It is true, with a slight twist. The sun will indeed move out of its orbit, gradually shifting outward, and eventually explode.

– If we don't have a sun, then everything will die.

– That is true as well, but there is always a plan in place when a sun moves, which involves an exchange of solar energy. A new sun will eventually enter our solar system in the future. This will not happen in the near future but is a process that will take a long time. However, it will come and take its place before our current sun burns out.

– So Earth will be okay?

– Yes, but by then, we will be living a completely different life than we do today. Therefore, we will also need an entirely different solar system. Thus, an exchange of solar energy will be necessary.

This is indeed a distant future event and not something we need to address now. But this exchange needs to happen for our solar system to survive. At the same time, other planets will also change their positions, as they will be affected by this shift in energies. When one part of a solar system changes position, other parts must adjust to follow the new solar system being formed at that time.

– Is this when the new part of Earth will take place?

– Yes, but the new Earth is a shift that will occur much sooner and is already underway. As you mentioned, it will be part of the same future shift. We will be living in a completely different way than we do today. The sun's power will then provide us with new energy channels.

– Why is it important to include this event in our book now?

– It is important because a change in the sun's orbit will occur in the near future. We want to explain why this change is necessary so that you won't be worried. It is a natural process where the new sun will be part of the new world we will live in.

– Is the new sun already on its way?

– It actually is. Although it is very far off, the process has already begun. So if you see a sign of a sun somewhere in the universe through a telescope, that is the one that will be part of our new solar system.

– Can we really see that far away?

– Some can, but we are talking about very advanced telescopes.

– How can the sun move and how can our solar system change formation?

– It will happen through gravity and magnetic fields. But there is also another force in the universe that humanity does not yet know about, which surrounds its power in a circumferential manner, leading to its movement. This is also what will illuminate your future.

But don't worry. Our solar system will always remain, but in the future, it will change its formation.

– If the sun disappears and the new sun hasn't arrived yet, what happens to Earth?

– We will enter a new ice age. It will be during that period when our solar system changes its formation.

– What happens to the moon then?

– It will remain but not as Earth's moon. It will also be replaced. The moon will change its orbit and will move to a smaller planet, which you call Pluto. Pluto will gain a larger place on your map in the future. Not because of its size but because of the power Pluto has within it, which will attract our current moon. Earth will then get a new moon with a completely different power that can work with the new sun.

– Why is this information important now?

– It is important because Pluto will indeed be reclassified as a planet due to its internal power, which you do not yet understand. Pluto will bring new knowledge to you. The reason you haven't reached this knowledge yet is that you have dismissed it as a non-useful planet. But for us, a planet is not defined by its circumference but by its internal properties, which you will become more aware of in the future.

This is very advanced information and naturally cannot be proven. You will have to either trust our words … or not. We cannot just write about things that people will believe or already know; otherwise, there is no development. Remember, if our solar system could be created once, why couldn't it change.

NEW SOLAR POWER

Kazandra speaks:

When the sun enters a period of change, which you will also experience, you will no longer be able to measure its power in amperes. This is what we mentioned earlier. You can only do that at the beginning. So even though those of you living on Earth today won't be affected by this change in the sun, the change will start its journey right now. This is what we mean by the time of change. Therefore, it is crucial that you start adapting now, as the sun will change year by year. The power that changes then is the very emission of power that you will use in the future.

– Like a power booster, then? I ask Kazandra.

– Yes, exactly. When power needs to increase, you will use this new power as a booster, and as you said: increase the power. When you do this, you will no longer be able to calculate it mathematically, as the power it will possess will be entirely different from what you have today.

– How is it that the inner power of the sun will change?

– This always happens when the sun begins a so-called shutdown. Therefore, you will no longer benefit from the same energy from the sun as you do today because that energy will be different.

– Is this when we create our own version of solar magnetism?

– Yes, because while you are working with your magnetic abilities, you will also discover that you will no longer need to rely on the sun's power, as this will be a process that will manage itself once you are finished. Just like we have a solar workshop on our ship.

– Won't this affect both humans and Mother Earth if the sun's power diminishes?

– It will become colder for a time, but not as you think. Your poles, as your angels mentioned earlier, will switch places. Where it is warm now, cold will prevail, which will be a welcome change for that part of the world and vice versa. So it will affect you, but it is a change that will take a long time. So do not worry.

However, we mean that the power you need from your solar energy, especially as today's sun wanes in strength and before the sun is replaced, will not be provided by the sun as needed in the future.

– Can't we use our solar panels then?

– You can, but you need, as we mentioned earlier, the full range of the sun's power. This is the part we want to address now. Because when you reach that point, and you notice that the sun's internal power no longer provides the same heat-retaining effect, it's not the end. All you need to do then is to get closer to reach the power you need. This is what we mean when we talk about capturing the sun's closest rays before they wane, as they will provide you with new power.

– Why those rays in particular?

– Because they are much stronger and more durable than you think. So start by studying the sun's beginnings and its entire range. Furthermore, it is essential that you also illuminate the internal concept of the sun's range and multiply it with the external range of power. Otherwise, you cannot increase the power the sun brings from the inside. How do we do that? You do that, as we mentioned earlier, by better understanding the sun's interior, and when you do, your colorful pattern will also become more apparent.

Solar magnetism and solar magnesium will
become the future's teachings (electromagnetism)
through the sun's energy waves and
their manner of travel.

The angels conclude:

We are pleased that some researchers have finally taken on the study of the sun. You will soon receive the information you need to understand how the sun can act as a source for Earth in the future. Therefore, it is important that you first learn more about this gigantic star, in terms of how we can help Mother Earth. Not how we can learn about other parts of the universe. We will address that later.

– Why has there not been more research on the sun? I ask the angels.

– It is a fairly simple question to answer. Humanity's hunger was in a gigantic gap where we differentiated ourselves between lands and peoples. Instead of understanding, humanity competed with each other about who got there first and paid no attention to how we could use the knowledge we reached together.

But now that research has taken on a mission, we have begun to use the information we receive in a completely different way than before. We are learning to cooperate and what is significant, both for our own and Mother Earth's survival. Therefore, you are also beginning to be more interested in the sun's range, and the next step, as mentioned, is a solar pattern. This is where you will gain new knowledge and learn to calculate anew. Therefore, we will now let the sun's mathematical union conclude this chapter.

THE SUN'S MATHEMATICAL UNION

The sun's mathematical union speaks:

According to the formula that includes solar energy, we can expand both the inner and outer parts – the larger and smaller segments. It's through a range of different combinations that we can reach both our internal and external power, and this is often divided into two. This is where the "divide by" button comes in and where it all started. By learning about both the sun's internal and external strength, we will, with the help of the sun, achieve new and better mathematical calculations. An inflow is, in fact, controlled by the sun, and through the knowledge of calculation, we will reach new and improved solutions in the future.

The mathematical concept will thus change its structure in the future. This is because we need better and more advanced calculations to use even more sophisticated formulas. The ones available today are insufficient, which has prevented humanity from achieving its goals despite many previous attempts. But once you have access to new tools and a new method for calculation, you will understand.

By being able to calculate the sun's phenomenal properties, you will also be able to calculate and benefit from the sun's range to improve your way of life. This also includes protection that humanity can use. Through a new and improved mathematical technique, you will also be able to calculate and live in a free world where everything is interconnected.

This is a mathematical source of the sun that you will learn about from your professors. It will come before we change the year to 3.

So embrace these new formulas, which come from new discoveries, and may initially be surprising. But every beginning can lead you forward, where everyone can live within the sun's range and calculate in the same way.

Additionally, some professors will change how you approach the sacred power of the universe. This is significant because, at a suitable pace, we will introduce you to new teachers needed for the future. So, learn from those who think differently and in new ways. Move on from the old and those who remain stuck in the same mindset. A new era in the mathematical genre is already on its way. Embrace this new, positive change and learn to calculate anew.

Both the sun and the moon will thus play a significant role in these calculations, as they are the focus of our attention both now and forever. It is not about what 1 + 1 equals. Numbers will no longer be as important, as future formulas will rely more on letters.

But teach children all the content, both the significance of numbers and letters, and you will see that our little stars will soon grow into the great scientists they are destined to be on Earth. They will then continue where Einstein left off in his career. A new extension of his theory of relativity will illuminate your beautiful world again and explain this concept further, but in a different way than before.

An as-yet incomplete formula, but one that will soon highlight a new rich source for you – $E=mc^3$. An ending here will be necessary and will need further development, as you are missing yet another "=".

The angels conclude:

For the sun's rays to travel, they need energy. This energy is what the entire universe is made of – a single energy flow, a single energy system. And this system is used by both the sun and the moon. It acts like a sort of transport route. So, one could say they are hitching a ride but are created by their own power, and it is what happens during this journey that interests us. Therefore, we will now continue with the light.

CHAPTER 3

LIGHT

INTRODUCTION

Archangel Metatron speaks:

To understand how we can use the energy emitted by today's sun, we need to start by grasping how light functions. Additionally, to comprehend light and integrate it into our solar magnetism, we must first understand concepts like spectrum, color, and tones. These elements are key to surpassing the speed of light.

Let's use a candle as an example:

For a candle to burn, it requires a wick, but for the wick to ignite, we need something to light it – such as a match. However, since matches don't exist in the universe, the universe has developed its own ignition mechanism, which, in this case, is motion. Believe it or not, everything can be initiated as long as there is movement.

What kind of movement are we referring to that can ignite both our own lamp and the universe's? It's core movement. Core movement is what orbits and creates energy waves in the universe, and it is what initiates the universe's light. Without movement, the universe cannot generate waves, and without waves, the sun's rays would not reach the Earth's center (we will discuss core matter later, where core movement is generated).

Moreover, for a flame to continue burning, it needs oxygen; otherwise, the light would extinguish. Since the universe does not contain oxygen in the same way that you can breathe it, it needs something to sustain the sun's energy waves. This is where our spectrum comes into play, as it is one of the things that maintains the universe's light.

SPECTRUM

Archangel Metatron speaks:

As previously mentioned, the sun's energy waves travel at an enormous speed, and it is at this speed that the sun's rays receive a new flow of energy. It's like picking up passengers along the way, but in this case, these passengers add new and powerful energy that the sun uses to increase its strength and reach even farther. Some of these passengers are ions, which isn't surprising since the sun's core also consists of ions.

– Earlier, you mentioned that the moon helps slow down the sun's energy waves to prevent them from becoming too intense for humans. But now, we should add elements, such as ions, to increase their strength, right? I ask Metatron.

– Good that you bring this up. We understand that it can be confusing. What we mean is that if the moon hadn't balanced the energy waves sent out by the sun, you would encounter entirely different forces at a completely different speed than you do today. But in this case, when we aim to create a new and improved power, we need to use the same technique but in a different context. We need to learn how to exceed the speed of light.

How can we apply these ions to our light? And how do we apply a spectrum according to our thinking about the path of light? But first and foremost, what is a spectrum? Many are familiar with how a spectrum works when light splits into different wavelengths. But how many know that we can also use the same principle to divide other electromagnetic waves based on their speed? And what really happens between these segments, in the vacuum of space where light moves and exceeds the speed of light?

– What is a vacuum?

– A vacuum is an empty space. An empty space that really shouldn't exist, but since it has a name and a presence, it does exist. In a vacuum, energy is stored. It's a process where we store and transform various forms of matter. Thus, humans perceive it as a division of light, when in reality, it's a transformation of light. This transformation occurs in various stages depending on the substances light passes through before being divided. This process can be used for our needs.

A spectrum, therefore, consists of several layers of light after a division has occurred, where you can remake and manufacture new things from what comes in – essentially a converter. This is exactly what we need now, but improved, a converter that can transform both the sun's and the moon's energy-rich rays into the energy we need for the future.

– I thought a spectrum was, as you said, when light splits into different wavelengths.

– Yes, exactly, and it's that division we want to use. But it's not just the process itself we are after – we also want to increase the strength.

When light is in a vacuum in space, that's when we reach the speed of light, because there's nothing there to slow it down. But when light reaches water or sand on Earth, it cannot travel as fast because the particles in the water or sand slow it down.

What we advocate instead is the vacuum in space, where light reaches its peak. It's when light reaches its peak, and there are no other factors to break or slow down the light's speed, that we reach the speed we want you to use. But instead, we add other components that can increase the strength of light.

This is where we also come to refueling, which we do when we pick up passengers. Each refueling allows us to increase the speed of light, and if we do this, we can also apply the same process in our electromagnetic field. That's why we call it a spectrum, because only after a division can we add other energy-boosting components to increase speed, and that's exactly what we're aiming for as a power source.

We should also mention that many researchers have already gathered a lot of information about the early journey of light. Therefore, we felt that information was not crucial for us; instead, we focus on the process that occurs after a division.

– What should we place in the vacuum after a division has occurred? And how will we know what to transform it into?

– You will find out soon. Here we just want to highlight its importance, and we believe this is a good path to achieving better electromagnetic solar energy.

– So we first need to create a vacuum. How do we do that?

– We cannot create the same vacuum as exists in the universe. A vacuum is something that only exists or arises on its own.

Therefore, we need to create something similar to a vacuum, where nothing else exists and where we can transform the structure.

– How can we both add and transform particles if they are the ones that slow down the speed? How can we then increase the speed or use it further within the electromagnetic field?

– It is essential to find a good solution for this vacuum and create components that are beneficial for the environment. We cannot create anything that slows down the speed. Therefore, it is important that we don't add any substances until after a division, as there will then be more rays we can use and transform anew.

This is also where we can improve our understanding of the speed of light and better grasp the process occurring in a vacuum.

THE SPEED OF LIGHT

Angel Sepatzon speaks:

We angels have always been able to move much faster than you can blink, and if something can move faster than you can blink, it can also surpass the speed of light. Therefore, we challenge some scientific theories that claim light cannot be broken, that the light barrier can never be completely separated, because if you can never reach it, you can never surpass it. And if you cannot surpass it, then you are not faster than it.

When we talk about the speed of light, we always include ourselves because our thoughts affect the communication we have with you on Earth. As a result, some may find our energy flow very confusing because it is both fast externally and strong in its nature.

– How does this affect us when we write automatic writing? I can't write at the speed of light. I ask Sepatzon.

– You are delightful. We don't mean it that way. When we pause in your channel, we can hover and visit you. So when we float and move with the power of thought, it affects both us and you. But we can also divide ourselves and pass by quickly without you noticing. Although some can, but only those who have learned to embrace the small, and it is in the small that we find the speed of light.

By "small," we mean that every person, object, or other form is created from a single tiny atom. So if you understand how these small parts work and are built up, you can also learn about the larger scale. Everything large always originates from the small, the simple, the part on which we then build everything. If we don't first learn and understand the foundation, the small, we will not progress, either in our spiritual work or in our understanding of the universe.

– Excuse me, but what does this have to do with the speed of light, dear Sepatzon?

– Even light began as a single tiny atom, which then grew larger and larger as other phenomena and particles were added. So if we don't understand how a tiny atom is constructed, we cannot understand the further construction.

There is a margin of error here that humanity has not accounted for, which is both the mass and the particles traveling at the speed of light. It is by adding material along the way that we can reach the same height. Otherwise, we would not be able to travel the same path or achieve the same results. Therefore, many believe we need to become faster to achieve more, which is not true. What we need to do instead is to outsmart its nature.

It is through each accumulation of new material along the way that speed increases. It's like refueling at every gas station. But a car, despite all the gasoline, cannot go faster than its capacity because there are limitations. However, in the universe, there are no limitations, so light can also accumulate anew. So with each new addition of light, the flow of light increases, and with each increase in the flow of light, the speed of light also increases.

Therefore, we will break the barrier in the future that prevents light from going faster. And this is exactly what we angels do; otherwise, we wouldn't be able to reach so many simultaneously. Thus, we can exist both in the spaces of the future and the past. For we too live and operate according to the speed of light because everything around us does. Thus, what negatively affects the universe also negatively affects us, and so on.

For this reason, it is crucial for you to learn more about this and visit more laboratories where you can collectively learn how to break the speed of light. This is an improved technique already used by other solar systems to great effect. Through this technique, they have already achieved several advantages because they have a better understanding of themselves and the universe as a whole.

A new lighter future will come where we reach beyond and surpass the speed of light.

Slow down – and we can reach
the spaces of the past.
Speed up – and we can reach
the content of the future.

Humanity says that when light has reached its peak, it is fully developed, but this is not the case for us. There is always a continuation, even in this area. And to understand it, we also need to delve into colors and tones.

COLOR SPECTRUM

The angels speak:

When the sun's energy waves reach Earth, they spread their energy (particles) across the entire planet in the form of heat, energy, and colors in their most intense form. Therefore, we will now discuss what these colors are and their significance for both us and the Earth. Even the colors of the rainbow are influenced by these particles (ions) when the sun's rays reach our atmosphere. And for these particles to move, they need something to cling to, and in this case, they cling to the sun's gravitational pull.

– What does this have to do with the rainbow? I ask the angels.

– A rainbow is a spectrum of light, and we want to focus on the effect of the spectrum because it will reveal its hidden knowledge in the future. There are actually more color spectrums that humanity has not yet observed.

– I have learned that a rainbow occurs when the sun shines on falling rain, and that is when we can see and experience a spectrum of colors.

– That is correct; it's a beautiful display of colors that happens when the sun's rays hit the droplets. But that is not the connection we want to convey now, as you already understand that aspect. What we want to convey is a different approach – how the sun, with the help of its passengers, can also influence our colors on

Earth. For the sun also has a significant impact on what happens when colors light up the Earth, and it is this colorful phenomenon that we want you to further develop now.

Using ions as a future example is based on ionized energy levels and associated colors because this will increasingly be used in the future. And this is where light comes in. Therefore, we want you to expand your knowledge of how ions, together with the sun, interact with the water that the Earth is composed of and how this interaction can create more beautiful displays of colors. When ions are exposed in water with the help of the sun, they shine in their brightest colors.

However, because the ions of the universe are more elusive, this is why humanity has not yet embarked on this fantastic journey. We have not yet embraced it in our electromagnetic sphere, but we know that you will do so in the future. Therefore, coloration and understanding are also significant when we ground ourselves in future renewable energies because different colors symbolize how far light can travel. When we understand this color spectrum, we can use it in our plan for various future energy levels.

Think of it like a traffic light where the color changes with each movement. It starts with red at the top, then intensifies to yellow, and when green, you drive away. That's when the strength is at its peak.

So study further into what can be transformed with the help of the sun's range, just as the rainbow does. This is the part of the spectrum division we are interested in and how, by embracing ions into this world of light, we can also create other color spectrums. Why wouldn't the addition of ions cause our light to both expand and evolve? Especially when the air, which merges with the sun's rays, already contains both positive and negative ions.

It is also well-known that negative ions make people feel good, so we need them. Therefore, if something negative in this sense can make us transform both our inner and outer well-being into something good, why couldn't an ion integrate into this world of light and, with the help of the sun, permanently change the color spectrum we use today?

TONES

Archangel Michael speaks:

If we can expand the light and make it spread in all possible directions, we can also understand the light signals that the universe sends out and where they come from. For if we don't learn how an instrument works, we can't play it. But if we understand where all the notes and tones come from, it becomes easier to play the tones of the universe. For every division that light makes, a new sound, a new tone, always arises. And for every photon that the soul is touched by, if you are attentive enough, you can reach these sounds. Tones move in waves because light does. Thus, they achieve a fine cooperation.

For example, if we strike different glasses with varying amounts of water in them, we have learned that they emit different tones. But what happens if we expose them to the light of the universe during the same ongoing process? Would the light passing through then change the tones depending on the amount of water?

We know that this is a method that already exists among you. But what we mean by this example is how humanity works to achieve a new result. Therefore, we want you to focus on the light and sound of the universe using the same technique as the glasses.

The glass can be a star. Can we hear the sound of a star when other light touches its exterior? And what determines how a star sound? Is there light and sound between the stars? What happens if a star is full of hydrogen oxide; does it produce a different sound then? And does it differ when the light comes from other parts of the universe, and so on?

This ability to hear will also increase in many souls in the future, where we can capture the sounds of the universe through our own sensors. Therefore, some will struggle for a while before they understand what these sounds are since we are not used to the sounds of the universe because our 3D world has blocked these wavelengths.

So if you suddenly start hearing strong tones with vibrations in the background, it may well be that you are beginning to sense the universe and the power that exists there. Then you can come into contact with that specific star. Ulrika already has this ability to hear and lives with it daily, and as mentioned, more and more will expand their abilities in this area.

Every point of light in the universe
always consists of a tone.

There are already many highly sensitive souls, and more will come in the future, who can perceive these tones, and they are the ones who will further research this. We need researchers who are receptive to the inner journey we are on now, as this is the only way we can understand the connection between light and tones. They need to hear and perceive these tones through their souls to integrate them. In this way, they can unite them with the light they see within themselves.

The entire universe is created from different light stations, each with a specific tone.

Those with advanced spiritual abilities can find and connect these tones, which often happens because you are at that specific light station in your spiritual development, and you may need to work more on it. Or you simply feel more at home in that tone because the soul always reacts to what it recognizes, which often depends on the teaching you are in at the time. And it is this ability to sense that we need more of in the future when we create more light without involving electricity.

That is why we have chosen to spread this knowledge now and let our young souls who research the light of the universe do so. So a good idea is to find the tone that is significant for you in your life. Your soul will feel it deeply within you and understand the light source we all come from. So if you see and understand the connection between how light can expand and at the same time reach the tones of the universe, then your knowledge increases. For if we, like an atom, understand that it cannot only be split but also learn what happens around it, then we also gain more knowledge. The same applies to light.

This is how we want to link future work methods, even for research purposes. Therefore, we now need more researchers who can understand and reach light and tones to understand how the universe works. For if you cannot sense this touch, it will be difficult for you to achieve future light results.

Brief Summary:

In a vacuum, energy is stored. When the sun's rays reach this vacuum, a division of the light occurs. We then add particles but only after a division. After that, another division occurs, and we can add new particles and so on. The more we divide and add particles, the faster the speed increases, and we can surpass the speed of light.

When you understand what each color symbolizes and how far the light can travel, you can also adopt it in your future color scheme. When you understand how light spreads or extinguishes and that the tone does the same, you can use light signals in different tuning systems.

When you understand the importance of the universe's interaction, you will also develop yourself. For it is a collaboration that must happen now for change to occur. We must learn how to integrate with the universe and see ourselves as the integral part we actually are. A piece in the puzzle that the universe is created from. We are not the ones who control it but a part, a small but important part of everything that governs, teaches, lives, and exists in the universe.

CHAPTER 4

THE UNIVERSE

INTRODUCTION

Nikodemus speaks:

Now, My Beloved Friends, we will finally drift among stars, black holes, magnetic fields, and a lot of other fun things. We have divided everything into five parts, and these are the forces we will discuss up until the next section. The purpose is not just to explain how the universe works according to us but also to highlight the parts we will carry with us into the new Earth.

1. Who created the Universe?

2. The background of everything – our three producers. These are the ones who made it possible for everything to take place and grow.

3. Gravity and magnetic fields, which were necessary for our first star to form.

4. The importance and development of stars – for the continuation of the universe.

5. Black Holes and Quasars.

Let's begin our journey by exploring each part in detail:

WHO CREATED THE UNIVERSE?

Nikodemus speaks:

The universe is everything, including our part of the spirit world, because we and the universe are always integrated with each other. We are all ONE. But there are rules that even the universe must follow, despite having no boundaries, and it is precisely these rules that humanity needs to study now. The universe's main rule is: everything we touch, we always give back. Everything the universe touches, the universe always gives back. This is a sacred scripture and it is always by this that we live and work in the spirit world.

If we do not act in accordance with the laws
of the universe, we cannot live there.
If we do not follow the rules that apply there,
we cannot achieve the same results.

But where does the universe's sacred power come from in the first place? And why is it so difficult to predict that we sometimes receive incorrect information? The answer to the first question is God. It was God, our Lord, who was involved in creating the first assumption of stars. Therefore, the universe's power is so sacred.

The sacredness of the universe's power, for both us and humanity, stems from the holy power that God possesses – a star's both inner and outer strength. Without God's holy power, which he gave when our first star was created, it could not withstand the enormous power that the universe consists of today.

God's holy power is so powerful that he can maintain the entire universe, and he always gives a small piece of himself to the stars that are born and collects the dust that forms when a star die. Thus, nothing goes to waste. Everything is used to create anew. And the more his talent increased, the more the universe's power grew. The stronger a star became and developed, the stronger God's holy power became, which he then used to help humanity advance.

– Does God stand higher than the power of the universe? I ask my father.

– No, we don't mean that. We cooperate and always follow the concept and the flow that exist in the universe. Therefore, the soul never suffers and always has a home to return to.

That is why it is called the universe's sacred power, and God uses this power only for good.

THE BEGINNING OF THE UNIVERSE

The angels speak:

So, how did everything begin? How did the universe start? And what came first, the chicken or the egg? Moreover, how do we figure out what came first? And why is it so important to know? Or rather, do we need to know what came first and, if so, who says that we do? And can humans continue researching even if we never reach an early result about what came first?

So, if we say the chicken came first, because without a chicken we can't get an egg, you might say: "But without an egg, we can't get a chicken!" See how we can twist and turn everything. Therefore, we instead introduced a beginning, the universe's beginning, and it is always from that point that we collaborate.

It's by understanding and accepting this that it becomes easier to grasp the fundamental issue – a primary result. But to understand the fundamental issue, we first need to understand the function. In this matter, we cannot begin with a genesis until you have learned how a star is created, which you already know. It is only after this knowledge that we can start to comprehend the beginning of the universe.

The beginning involves learning about *how everything started*, but also *how everything then developed*. It's like a reverse film, where you first create an event and then go back to how it all began. So, you already know that gas is required to create a star, but to get the right gas, we also need to add other components, including one we call *reaction*. But where does this reaction come from? This is where we arrive when discussing what came first – the chicken or the egg. We go back to how everything began.

We have divided the beginning into two parts:

The Beginning Part 1

Many already know how to split an atom, which involves sending a neutron at a uranium atom, splitting the nucleus and releasing new neutrons. These, in turn, can split more atoms, creating a chain reaction. But historically, this splitting (chain reaction) didn't exist from the start; it was created as a singularity from the very beginning, as a standalone reaction.

Does this seem impossible? To us, it's the most obvious thing in the world, because if a single reaction couldn't act alone and reproduce, there would be no development. Therefore, we need to accept that a solitary event, a small part of the universe, could grow despite being independent of others.

This isn't a new phenomenon either. There are cells on Earth that can also expand their circle on their own. Everything happens according to the gravitational pull of the Earth, the sun, and the moon, and it's important for humans to understand that.

– How could it grow? It would be like a woman becoming pregnant without a man's sperm. I ask the angels.

– You are absolutely right, and that's actually how everything started – that a solitary reaction could produce more reactions on its own because it was created to do so. There was an analogue at that time, a signal, before the Big Bang took place. It was this that then had a certain influence on the same reaction.

Furthermore, don't forget that the universe looked completely different back then compared to now. Therefore, the substances acted according to what was present at that time. Everything started with a single thing that could double itself in number. Though with some influence from other modest things like a reaction.

– But then it wasn't alone, right?

– It was. An analogue is a signal while a reaction is created by an event – something happening. They don't interact with each other; one arose from the effect of the other and that's how it developed.

– So analogue and reaction are like the chicken and the egg – who came first?

– Yes, exactly. But the difference is that we know the reaction came first, but it was the analogue that then further developed and created the universe. It was the analogue reacting to the reaction that then sent out more signals and thus doubled them.

This is the first part of the universe's beginning, that a single singularity could double itself. We just need to accept that it was so because the universe looked completely different at that time.

The Beginning part 2

For a solitary reaction to act and double itself, it needed to consist of something strong. A weak composition with just, for example, water molecules wouldn't emit the same strength in its solitude. What was required was something that could form gas, including nitrogen, helium, and so on. In the atmosphere that existed then, there was only a gigantic darkness because the first star had not yet been created.

But as many already know, life has been found even in the darkest of water holes. It's possible to find something even in the most inhospitable, dark space. This is also where dark matter comes in, which we call core matter because it was from this that everything then took off (we will discuss core matter further later).

We now have reaction – analogue – gas (hydrogen and helium). The process was underway.

This was what formed the foundation:

- That alone soldier could reproduce.

- Finding the original source, which we find in core matter.

– What does this have to do with stars, my dear ones?

– It is very important that we always first understand the foundation, where everything comes from. But you will never get a hundred percent answer to that question. We only mention core matter and a lone soldier because they are the beginnings of everything that came later – how gas and a star could form. Core matter was just a space where all newly created energy gathered.

– Didn't the spirit world exist at the same time?

– No, the universe is much older than we are, but we didn't come much later. The soul got its own time to be created when our stars were created because the soul couldn't be created until we had reached the substances that a star contains, what the soul is made of. At that time, there was mostly floating matter, so it also took a long time to develop a soul.

The development of the soul has therefore always followed the development of the universe to adapt to the conditions that prevail there. When life then began to take place on Earth, the development of the soul also increased. This was to give it the conditions it needed for its own survival, and as humans were created and developed, the soul did too.

Summary:

- The reaction started everything.

- The analogue was created from the lone soldier's reaction ability – the reaction created various signaling systems. The analogue, in turn, had some influence on the reaction, which in turn created more reactions, and so it continued. They simply started to interact.

- After that, other elements took over and created a process, which we now call *the universal process*, allowing the universe to begin its journey and our first star to form. This is where core matter comes in. Core matter was thus created by and in the process.

We have drawn a simplified picture:

How they began to interact and expanded core matter, which then together created a core reaction.

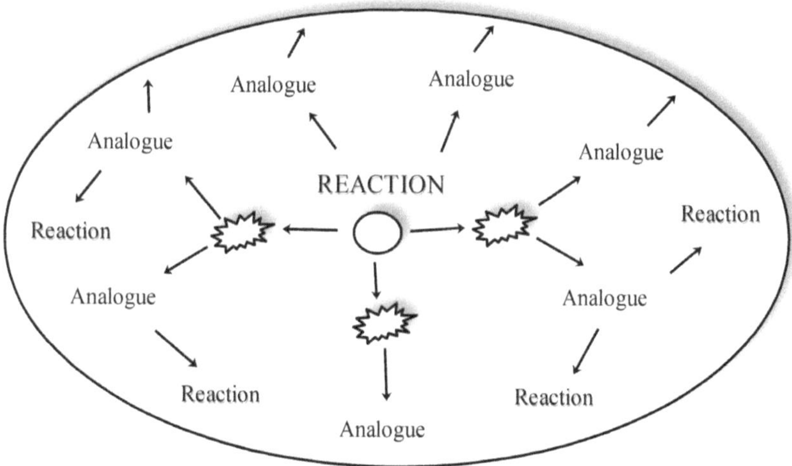

– So if the analogue hadn't been created, what would have happened to the reaction?

– Then the reaction wouldn't have been able to *assert its power*, as we call it, and create a continuation.

– Then the reaction needed something else for its continuation. Is it still considered a lone soldier then?

– It is, because the reaction was created in its singularity; it could double itself without any other help, and this is what we mean by a lone soldier being able to reproduce. But naturally, you're right, everything continued once the analogue took hold and could then start the process.

So all in all, a reaction was needed that created signals, analogues, that could increase its power, and the process was underway, which eventually created a core reaction that you call The Big Bang. But it wasn't until long after this that The Big Bang occurred as a result of the universe's expansion. The Big Bang also didn't happen exactly as described in your books. Yes, an explosion occurred, but not until much later than you are aware of.

This is not something we will go through now, but The Big Bang actually has a completely different background than you know. That's why we call it a core reaction instead. The beginning that arose then, before The Big Bang, will be written about by other highly spiritual souls, who will tell the correct version of how and why the universe's power arose. Parts regarding The Big Bang are, in fact, a fabricated version, as they didn't have another explanation available at that time, and they wanted to release their version long before the correct version came out.

BOUNDARY

The angels speak:

The foundation for the universe's vast development lies in the freedom it possesses. It is not governed by any boundaries. This lack of limitations is crucial; without it, we wouldn't have the universe as we know it today. The universe thrives on its limitless nature. It is humans who impose limits.

How can we make such a bold statement? We can because the universe is everlasting – what exists within it can never cease to exist. In the spirit world, we live according to the universe's energy flow and laws. If there were a finite space in the universe, it would also limit your soul, including the part involving photons, the aspect that gives the soul light. If both stars and photons were to disappear, the soul would also vanish. Hence, the universe is created as one vast, hollow flow where there are no limitations, and everything can be recreated.

For example, let's say Mother Earth's resources were to run out, which would have devastating consequences … for humans, not for Mother Earth. Mother Earth can always regenerate over time. Therefore, humans have limitations because you depend on the Earth's existence and functionality; otherwise, your physical body will die.

So, humanity's critical limitation lies in the health of the Earth. If we do not care for Mother Earth, we also lose the boundary we had before, thereby shortening our lives – we diminish our limit and our destiny. But the universe doesn't work that way.

Everything in the universe can always be recreated, and if it always exists and can always be recreated, then there are no limits.

Therefore, it is crucial for humans to learn to harness the universe's power as a future source of energy because the power there can never run out. But if Mother Earth's oil runs out, it's gone. If Mother Earth's coal runs out, it's gone. If Mother Earth's sand runs out, it's gone. Nothing can be recreated. It is crucial to decide which branch you want to belong to, the limiting world or the limitless one.

– How is it that everything in the universe can be recreated? I ask the angels.

– The ability to recreate everything stems from a foundation. If there is always a foundation, it can never end. That is why the analogue and the reaction were so important because they are two elements you can never remove.

The universe is also very frugal, constantly ensuring the foundation remains. It saves, it plans, and thus continuously creates a good starting point for itself. In contrast, humans have not yet reached this tactical capability because you dig and dig until even the foundation is gone, and if the foundation is gone, it can never be recreated.

This is the problem for humans. You hunger so much for more and more that you eventually even remove the foundation. You build and build and live in a world where you believe humanity is invincible, ignoring the warning signals and continuing as if nothing has happened. The universe does not operate in this manner. That's why the universe can never extinguish.

WILL THE UNIVERSE EVER CEASE TO EXIST?

The angels speak:

Of course, stars will always die, but nothing can ever be completely extinguished because the foundation of the universe is so strong. We do not buy into this negative narrative. It is merely a scare tactic meant to make you live in fear! The universe will NEVER die out! So don't worry. Some scientists do not possess the full growing power of the universe, and we should not discuss what we do not understand. Current researchers have only 1.5 % of the entire knowledge field of the universe. This is a bold claim from our perspective, considering their limited knowledge in the area (Your books say 4.5 %, but this is incorrect).

Therefore, we find it hard to understand why humans calculate as they do and claim to know when the entire universe will end. This is despite the fact that humans have only been on Earth for about 200,000 years, while the universe is billions of years old. But all calculations are a step forward, and somewhere we must start, and that is where humanity is, at the beginning of the entire field of universal knowledge.

So stop talking about the universe's demise because it will never happen! Do we have other universes to wander in? Yes! Do we have other stars if our own were to be replaced? Yes! Are there parallel worlds to Earth? Yes! Will our universe ever cease to exist? No! The universe can never disappear, extinguish, or die out. PERIOD! Because we have a clear foundation – rebirth.

It is based on this law that all living things operate and grow, that if something dies, something new must always grow in its place. So to understand how the universe works, we want you to talk about rebirth instead. And in the future, researchers will further develop this concept that the universe is infinite because rebirth is always stronger than the end.

Only the universe is eternally
a permanent part,
but its content will always
change its character.

The Angels Conclude:

To reach the beginning, we need to further research, utilize, and accept the universe's beginning. To accept that a single entity could multiply itself without external influence. And when we understand that the universe is an eternal source that can never extinguish, we can also use the same techniques that the universe uses to create the conditions we need to move forward. And the power that existed then, when everything began, is found in the core matter, and that is where we are headed now. For it is here, in the fantastic world of core matter, that everything truly began.

CHAPTER 5

THE BACKGROUND OF EVERYTHING

ASTRAL JOURNEY

Ulrika speaks:

What exists out there, beyond everything? Through an astral journey I took with Archangel Metatron, I obtained the following information:

I sit down in my meditation corner and fill my inner self with relaxing music. I breathe and relax. I begin to float outside of my body. Suddenly, I find myself in the universe. I am in the part that humans call dark matter. The anatomy of stones surrounds me. These are the building blocks that make up the foundation where everything is created in the universe.

I see a lot of pink, a color that this part has given itself. I see stars, many in number, and all are part of the process occurring here. I float by and look around.

The color green becomes increasingly noticeable, washing around me. It's a lovely color and it's here to process, repair, and assemble everything that exists and is needed here.

I come upon blue. Here, electricity is created, all energy, and

I hear it crackling and popping as I float by. I hear the sound of energy when it gets into motion.

Yellow becomes more and more visible the further in I go. Yellow, which wasn't present before but has now begun its journey in this core space. Yellow wasn't present from the start because it couldn't fulfill its purpose then. But now it is strongly present in this new, future process and is here for humanity to understand that it exists and how it channels all its energy in the universe.

Yellow represents forward movement in the universe. The forward movement that happens even when fire burns, but where red is the strongest color of all. I do not see red and I do not know why, but it is absent here. Perhaps it is a journey I will undertake later in another astral travel, in another dimension. To further unite with what lies beyond this processing curtain.

I make my way back.

Every color in the universe has always
represented its sacred power.
Therefore, the same color can vary in its meaning,
depending on where we are at that moment.

Archangel Metatron explains:

For us angels, the universe has always been the same vast size. The difference lies in its internal scope and that its internal structure has grown in strength and character. You will also soon discover that it is not as dark as you might think. You just haven't yet reached the light, at least not as far in as you need to go to understand it.

Let me give you an example:

Imagine a furnished living room. Suddenly, the light goes out, and everything becomes dark. You might think that your living room is empty because you can no longer see it. This is similar to how it was for the universe in the beginning. The universe was just as vast in surface area then as it is today, but since everything was dark back then – like the light being out – we could not see anything. But does this mean that just because the universe is dark, it is empty? No, everything always consists of something, even if it is invisible to the naked eye.

So, what is out there? What did Ulrika see in this captivating world of ours, in this dark world that is actually light? This is where we come to our three producers. Without them, the universe cannot spread its power further.

1. Core Matter
2. Interstellar Space
3. The Transparent World

CORE MATTER

Archangel Metatron speaks:

Core matter, also known as dark matter, is the foundation of all our creation. Some of its content existed from the very beginning, while other parts have been formed over time. The term "dark" is used by humans due to a lack of knowledge about the subject, as humans are not yet able to measure the structural relationships that exist within it. However, if you look up the word "core," you will see that it is still used today.

The term "core" originates from this concept, though it has been applied to various human components such as weapons, power, and reactions, and so on. As you learn more about core matter, you will understand what we mean. Therefore, we use the correct term, core, not dark, because it is a small part that has survived since ancient times.

Core matter is a composition that enables the creation of other components in the universe. Without core matter, there would be no growth. Some also view core matter as threatening, but without it, neither we nor you would exist. Core matter is the primary and greatest cause of our existence.

But how did it all begin? Let's return to what we previously mentioned:

1. A reaction that started the journey and created conditions.

2. The reaction led to the creation of analogs in the world of core matter, resulting in the formation of a core reaction. Core matter existed before the reaction, and its existence is based on particles transforming and creating analogs.

The process was born.

As we mentioned earlier, the universe originally contained a reaction. This reaction then multiplied and was vital for creating core matter, which in turn created an analog, leading to the formation of gravitational and magnetic fields, allowing our first star to form.

So, core matter was initially just a space, a substance, that then created everything. As time passed, this space encountered other spaces. When they combined their inner and outer strength, their capacity increased. At that time, core matter was mostly composed of *core material*, as we call it. Unlike today's core, it was a solid mass where the name had a different meaning based on the connections made at that time. When core met core, an analog was formed. The core was thus so strong that it was a prerequisite for development to occur.

– So, there were signals in the universe before The Big Bang. Where did they come from? I ask Metatron.

– An analog, which is a signal, could send out signals to another analog. The signals existed in the core world, where they were created, through their own actions. It's a bit difficult to explain, but that's how it was.

– So core matter is not empty.

– No, and it never has been. Nothing can ever be empty. If it were, it would not exist. There is always something there. You may not be able to see it with the naked eye or understand its origin, but as long as it exists, it is never empty.

It was an early beginning,
and it set everything in motion.

How do we then understand and learn more about the world that many might not yet grasp? We always start with what we cannot see because it is often where we find the answers to many questions. After all, what we can see with the naked eye is usually something we can touch, count, or measure because it exists.

But how do we approach what we cannot see? And how can we know that something exists if we cannot see it? Since the universe is not created based on human abilities to live, see, or act, we need to look with different eyes than the ones we were born with. Everything in the universe is created according to its own abilities, while the human eye is created to see what exists and operates in your world.

There are other living beings with vision so advanced that they can perceive other parts of the universe in ways that the human eye cannot. Therefore, for you, if you cannot see it, it does not exist. But for us, whether something exists or not does not depend on sight. It always depends on how developed you are to perceive what exists but is not yet visible to you.

Just look at how the owl's vision allows it to see in the dark, which you cannot. That may help you understand what we mean by the dark world. You know it exists because you can measure gravity. Therefore, it is no longer abstract, but you still cannot see it.

Archangel Metatron concludes:

Even now, some researchers are trying to explain the end of the universe. But this time, it's core matter that is being blamed, despite the limited understanding of the subject, which we find puzzling. If core matter, according to some scientists, is supposed to cause the end of the universe because it is said to be expanding the space between our currently existing galaxies, why wouldn't it also be able to initiate everything? Why is core matter believed to be able to end something but not to start something?

Furthermore, why would core matter not have existed until 100,000 years after The Big Bang? According to our timeline, core matter has always existed, at least since the day everything was created. This is what we mentioned earlier. Not in the same amount as today, but a certain amount always had a small beginning before it could expand to its full effect.

Something continuously
replenishes its inner substance, which is why
its impact increases.

It was also core matter that laid the foundation for the nuclear energy that came later.

NUCLEAR ENERGY

Archangel Metatron speaks:

You might already know that core matter and nuclear energy are two distinct things but belong to the same category. But why have we chosen to separate them? And how can we learn and understand their concepts and associated worlds? This brings us to our first producer – nuclear energy. So what is nuclear energy?

- Core Matter is a substance containing many particles that generate electrical current. This current is then transmitted from various sources, originating from the dark world, and is what we will refer to as nuclear energy in the future.

- Nuclear Energy is the space that exists *between* matter. It is from this space that the mass collects the particles needed to generate electricity and energy. This is what core matter feeds on. The same type of energy exists in other parts of the universe. It is created in the dark world

> and then combines with other components to spread throughout the rest of the universe. This is where all the colors exist, as mentioned during our astral journey.

Thus, the core world is a producer of a beautiful energy flow – its primary role is to produce. It is a technique that you will also learn in the future.

Let's use a battery as an example:

Consider a car battery, or any battery. A battery's job is to function as a power source. In the core world, however, things operate a bit differently. There, things are produced that humans might consider dangerous if they fell into the wrong hands, which is why there are names like weapons, power, reactions, and so on.

But without this core, the universe wouldn't function. It would be akin to removing the car battery; the car simply wouldn't start again. And the universe, which is so powerful, needs a powerful battery. Therefore, we need something that produces and acts as a link to other parts of the universe.

– How can this dark world, which doesn't contain light, create an energy flow? I ask Metatron.

– Good question, because this is exactly what we mean when we talk about its existence, even if you can't see it. And yes, we can produce energy even in the darkest space. It is not entirely dark; it's humans who believe it is.

We who live and operate in the universe can both see and access that part of the energy flow that humans cannot. To see it, you need a better vision. This vision will develop soon, and it is important for us that you start mastering this aspect, especially the energy part. It is through this dark energy that you will harvest electricity in the future.

We are not authorized to disclose more about core matter and nuclear energy. It is up to humanity to strive to understand this aspect of the universe.

Create new eyes that can see,
and you will understand.
Create new thoughts that comprehend,
and you will see.

Ulrika: I am now writing down what I receive, even though I don't fully understand its meaning. However, I have a sense that these are important components from the core world that we can learn from.

°2 is a significant component in dark matter: It can produce lightning. Helium fills the clouds so that °2 can create lightning. Hydrogen increases the concentration, which results in a change, an improvement, in C^1.

THE INTERSTELLAR SPACE

The angels speak:

The next producer is interstellar space. This is the region between the stars. Researchers once believed this space was empty, but it never has been, and we are glad you have finally come to understand this. Like core matter, it often sets everything in motion.

What then constitutes interstellar space, and how can we apply what happens there to our own lives? Interstellar space is actually composed of more gas than you might ever comprehend, and it is this gas that leads to the emission of energy. Without this emission, no development would occur. Gas and emission are crucial for our future energy research. This knowledge and technology are what humanity needs to embrace now, as it will contribute to a better environment. Of course, there are other areas as well, but we start here.

Additionally, interstellar space also consists of what we call *zonders*, which are particles floating between time and space. Since early researchers did not understand the concept of time and space, they believed interstellar space was both permanent and empty, lacking both matter and a fluid, changeable nature. However, this is not the case – there is much that floats within this delightful space.

– Why the term "zond," and what is a zond? I ask the angels.

– The term "zond" originates from an event where two pieces of matter meet and then transform into something new. This event also gave rise to the name "zond." It is from this that the term "sond" emerged, referring to the instruments astronomers send into space to gather new information. A zond in interstellar space acts similarly to your sond – it collects matter as your sond collects information, serving as a means for both to expand. Hence, the name sond. We changed the name to avoid future confusion.

Our Zond = collects matter – expands.
Your Sond = collects information – expands.

Our zond is a mass formed after it collides with another mass. Since mass in interstellar space can float freely, we refer to it as *the free world.* It truly feels that way. Moreover, there is no other place or mass in the universe completely independent of others; they can act freely based on their own content. While a mass may choose to collide with others to gain companions, it is entirely free to act as it wishes.

– How can a mass choose to collide with another mass?

– It can choose because it is free to do so; no other determining factors exist there. The choice of a mass is based on a desire to expand. Humans cannot expand on their own either.

– Excuse me, but it almost sounds like the mass can think.

– We understand how it might seem, but that is not exactly what we mean. The mass's choice is based on the gravitational forces present throughout the universe. It may seem like it chooses, but it is the gravitational force that creates the desire to expand. This is also where the universe's first gravity emerged. When a mass increases in size, its main task, gravity collaborates with other masses, similar to how the Earth, Sun, and Moon interact within our solar system. However, in interstellar space, it does not have to do this, although it can if it wants. Being in the free world allows it to make its own decisions.

A mass does not always need an event to expand. It can also attract companions, like a moon. The gravitational force of the mass is so powerful that it can do this. Among other things, it uses the moon as a light source. Thus, interstellar space is not completely dark. It receives both light and nourishment from the energy waves of moons, just as we receive from our moon.

This is what we want you to study further now, as it will allow us to postpone that matter at your desks.

– How can we incorporate this mass with its accompanying moon into our thinking within our own solar system?

– You can do this by studying the content even more carefully. By doing so, researchers will find a better solution for future energy supply. And if you understand the impact that interstellar space has on the rest of the universe, you will also understand what needs to be changed in your lives, now and in the future.

The interstellar force can never act without the rest of the universe. Nothing can. No one can act in isolation, grow, and change. Except for the reaction we previously mentioned. But once the universe came together, it remained that way. Everything is always based on collaboration, including everything related to our own solar system.

Without the universe, Earth would collapse, as what we depend on for our survival is itself dependent on something else for its survival. Everything in the universe integrates with one another and is connected like a long, unbroken chain. This is how everything evolves, even within the free world.

– How can it be free if it can never act on its own?

– It is free. What we mean is that it is free to choose whom or what it wants to collaborate with, and when it wants to do so. The rest of the universe already consists of a well-planned map of how everything should work, which does not exist within interstellar space.

That is precisely where all change occurs; it is where we create new things because it is possible there. Think of it as a place where everything is stored, transformed, developed, and redirected. However, it can also, if the gravitational force is strong and necessary, draw power from the rest of the universe. For it is the gravitational force that determines whether it remains within its own space or moves outside of it. In this way, it is only free within its own environment.

– What happens if it moves outside the interstellar space?

– If it moves outside, it does so because there is a need, and that need is relevant for you to research further now. When you understand this, you will also learn how to manage the universe's energy flow, perhaps not all of it, but the fundamental knowledge will be there as a determining factor, from which everything else will expand.

– How can knowledge of the interstellar space teach us about our own environment?

– By understanding the universe's gravitational forces, even beyond our comfort zone, you will increase your understanding of how we can integrate with other planets within our own solar system.

We will further develop this when we address interstellar gravitational forces. Here, we only wish to emphasize the importance of increasing our knowledge so that we can apply the same mindset to both our daily lives and ongoing environmental projects. By doing so, we will also understand the energy flow and gravitational forces we are so dependent on, just as the free world does between the sun, the moon, and the Earth.

Interstellar clouds and explosions

The angels speak:

The most intriguing aspect for us is that it is free. Therefore, the interstellar space does not always behave like the rest of the universe.

Let's use child-rearing as an example:

We have two children raised in different ways. One child has stricter rules imposed by their parents, while the other has more freedom to act as they wish. You might understand that their behaviors in response to different events will differ both during their upbringing and in the future. However, if they gain knowledge and understanding of each other's different behavioral patterns, they can learn to cooperate in the future. This is similar to how the behavior of interstellar space differs from the rest of the universe.

So, even though the interstellar space has different "upbringing methods," it does not mean it cannot learn to integrate with others, just like the children. But for this to happen effectively, and before it reaches the rest of the universe, it first needs to release its own energies. Otherwise, conflicts could arise. Thus, the mass acts out its behavior before interacting with the rest of the universe. Since interstellar space mainly consists of gas and other particles, you might understand what happens when gas meets other reactive materials – it causes an explosion.

However, an explosion does not always have to be of a harmful nature. There can be much benefit derived from it, and this is exactly what we want you to study further. This is a technique, an explosion, that you will use in the future. When interstellar clouds integrate with the same gas, a chemical collision occurs. This chemical collision can be utilized on Earth as an energy source. But to achieve this, more scientific research is needed regarding this type of explosion. Otherwise, it could be dangerous if mishandled or if it falls into the wrong hands.

Interstellar clouds consist of light gases and other particles. It is a place where stars are formed when a cloud collapse. Gravity then pulls the matter together, igniting the core = explosion.

– How are clouds created? I ask the angels.

– The creation and capturing of matter by clouds primarily depend on the gravitational force we discussed earlier. It is always the gravitational pull of energy that governs the density, which allows clouds to create waves.

– How can a cloud create waves without the external force that governs and exists beyond interstellar space?

– The waves are created in a manner similar to how the Sun sends its energy waves to Earth. The principle is based on the same event where waves emerge from the energy flow, the gravitational pull, present there.

We can also compare these waves to those created by a jellyfish, where its thin, fluttering, surrounding tendrils move only because of the waves in the water. Clouds act in a similar way, but in the universe.

It has also come to our attention that current scientists are not entirely in agreement about how gas clouds form. Some suggest focusing on the shape and structure of clouds to understand how a gas cloud form. Others argue that a small cloud cannot create an explosion, and so on. We understand these considerations. Therefore, further study in this area is necessary.

But we can say this: For a gas cloud to explode, a substantial amount of cloud is required. The remaining parts then take cues from and learn from the larger clouds, which is where the concept of integration comes from. Thus, a significant energy impact is required, and it is this distribution of energy that we are interested in as a future research area.

Additionally, there are some star explosions occurring in interstellar space. These star explosions pertain only to the old universe and are not visible to humans. What you see are gas explosions in the new world, but in the old world, we also encounter star explosions. The occurrence of these explosions is more related to their composition than to their complete development. But the foundation for creation is present, and it is this foundation that explodes.

You will understand when you open up to the old part. You will also see this preliminary stage of star formation, which is only visible there. Thus, when you establish the link between the new and the old, you will see how both collaborate to produce what the universe needs. Both are interconnected chains of the universe's creative power. If gas and other particles had not integrated with each other, no development would occur. This integration is what we want you to study now – how everything integrates. By doing so, you can both save and transform our world.

It seems that some researchers are puzzled by how the universe is created and operates as it does. However, this confusion stems only from the old source from which we previously obtained our information. Therefore, it is crucial that you now enhance your techniques, as this will lead you deeper into the mysteries of the universe. You will receive new information that overturns the old. This is how all research functions, and it is also why we should not discuss overly advanced future concepts.

Thus, it is essential for humanity to live and learn according to how the universe lives and learns, not based on how humans have learned or wish to live and think. This is where the issue lies; you oppose how Earth is structured, created, and functions. If instead, we lived according to the universe's construction and followed its governing laws, we would not be living under the conditions we do today. The universe knows; it is a perfectly designed constellation that humanity is only beginning to understand, but we are making progress.

THE INTERSTELLAR GRAVITATIONAL FORCE

Archangel Metatron speaks:

As we mentioned before, there is indeed a gravitational force in interstellar space, but to understand what it contains, you first need to understand why it behaves as it does. It is already well-known, even on Earth, that the force present there does not exist anywhere else in the universe.

We mean that it does not contain some of the fundamental concepts that some might claim, such as it not existing at all – but it does. The interstellar gravitational force is as real as you are. The difference between you and it is that it can stretch far more than your spiritual body can. Therefore, you do not yet fully understand this unique force because you do not yet fully understand yourselves, but you will.

So why are we discussing the interstellar gravitational force? You will need to highlight this force in your future power plants (we will discuss this further later, including when we talk about neutrinos).

– How can we use it in a power plant? I ask Metatron.

– To unify forces in the future, more than you already do, you need various concepts that can collaborate. The interstellar gravitational force is willing to collaborate; that is why we encourage you to use it. It is willing to collaborate because it can be combined with almost everything. Think of it like a puzzle where all pieces synchronize with each other, even though their shapes are different. But remember, the force in interstellar space cannot add its own strength; if it runs out of power, you will need to create a new.

Thus, it cannot be extended like some other forces. That is why it can be applied to most things. However, it has no determinative power and can diminish if we do not continually provide it with something, which is what happens in the universe. Therefore, the interstellar gravitational force does not diminish in the universe.

We also want to emphasize, as the most enduring actor in gravitational force, the enormous potential it has to bring about change. This is why it will be so important in the future, as it is both willing to collaborate and can be transformed.

– How can we transform a gravitational force?

– You do so while stretching or pulling it. But pulling on a force only happens when we need to use its capacity for other purposes, such as changing it. Think of it as an additional power source that can help you bind together different *fragments* (clues). This is what the interstellar gravitational force will be used for: as a means to act, transform, and unify different forces so that you can create the power that your future power plants will rely on.

Since the force is so changeable, we can also create different block stations. By block stations, we mean the parts where we combine different forces and divide them into various blocks. Here, with the help of the interstellar force, they can unite and increase their power before we merge them. It will be crucial for you to do this.

– What happens if we don't first divide them into different stations?

– To manage the interstellar gravitational force, it needs to first increase its internal strength; otherwise, it could fail. Therefore, we first divide them into different stations. But once they have integrated with each other, in the strength they possess, they are secure in the power you will manage for future use.

– What do you mean by "manage" in this context?

– To ensure you always have access to the same power source, it is essential that there is always an available power source. Therefore, just as you do today, you will set up various systems where the power can be gathered in different block stations before being combined with the interstellar force. This way, the power never runs out.

It is also crucial that you always manage forces that can continue and continue, constantly creating new energy without many new additions. This will make the power both cheaper and more sustainable, leading to a better and cleaner society. However, forces like this should not be assigned to just one category or country; EVERYONE should have access to the same source.

THE TRANSPARENT WORLD

Archangel Metatron speaks:

The third and final producer is the transparent world. The reason there is a world in the universe that humans are not yet aware of is because you cannot see it, and you cannot see it because it lies behind all the clutter. To see it, you would need a better long-range telescope, such as a kaleidoscope. Not exactly, but close – more like a kaleidoscope with glasses. This makes it appear transparent to you. And as we mentioned before: if you cannot see it, you will not believe in its existence. But it does exist and is as real as the universe itself.

The reason it is called the transparent world is that this is how humans perceive it. It seems not to exist, but it does. It also lacks walls and has neither a beginning nor an end.

It is a space that is not easily
accessible to everyone, but it is still there and can
express all its love behind the scenes.

To renew our thinking, we also need to renew our knowledge, and that is what we want to convey now. However, we will not go into how it came into being; here we only want to highlight that it exists and cover the parts that are important for human research moving forward.

The transparent world is a space that exclusively contains mass. It is also from here that we can reach out to the rest of the universe. It is the one that integrates with core matter and interstellar space, then processes and channels the matter further into the universe, where it is used for various purposes. Either to create new planets or for the stars to capture its matter. Compared to core matter and interstellar space, it is also a very high-level production channel, despite being the least visible.

It can be reached between core matter
and interstellar space.
It is behind everything but still visible in the middle.
Far enough away but still close enough.

Why have we chosen to highlight the transparent world in our book? There is always a reason behind everything we convey, and in this case, the reason is the connection that humanity has already achieved regarding core matter and the outer universe. By linking the transparent world with the outer universe, you can further develop the same concept in the same way.

The idea is that the more we learn about the universe as a whole, including what happens behind the scenes, the more we can learn and integrate it into our daily lives. When you do this and discover the mass that is there, the light will become clearer and clearer.

This is an informative source coming from God.

CHAPTER 6

GRAVITATION – MAGNETIC FIELDS

INTRODUCTION

The angels speak:

We have now covered the beginning of the universe and three of our major producers: those who created the gravitational forces, those who laid the foundation for gravitation, and the magnetic field we have today. This is also what we will delve into next, and everything is explained from our perspective.

We have divided everything into nine parts:

1. The entry of gravitation.

2. The small gravitation – The study of your body and soul.

3. Gravitation is not governed by time and space.
   When you understand the fundamental concepts of the soul and that gravitation is not governed by time and space, further future insights will emerge.

4. Gravitation is free to act in its own movement.

5. Gravitation cannot be controlled.
   Once you understand that gravitation is free and cannot be controlled, you will gain new knowledge that will advance your technology. However, despite being free in its own movement, gravitation can both cooperate with and be influenced by a magnetic field.

6. Gravitation as a stretching force.

7. Magnetic fields as determiners and balancers.

8. Gravitation brings things together.

9. The Earth's magnetic field.

THE ENTRY OF GRAVITATION

Archangel Metatron speaks:

Gravitation is crucial; without it, the universe wouldn't exist, and nothing would function, not even the tiniest atom. Everything would simply collapse.

– How and who created gravitation? I ask Metatron.

– Gravitation arose from an event, part of the analogy we previously mentioned. It was the force that initiated everything by pulling together matter and creating movement, leading to the formation of our first star.

Early scientists discovered how and why an apple fell to the ground. An apple couldn't fall without the influence of gravitation.

– Wasn't that gravity?

– Yes, exactly. Gravitation is also known as gravity. Gravity is the manifestation of gravitation's ability to move, existing between time and space. Without it, the apple wouldn't fall, and the universe wouldn't have formed its first star. It was a significant discovery at the time, hence we call it early. Understanding it within the study of the universe was crucial to developing the initial concepts correctly.

– How does the use of gravitation differ in the spirit world compared to us on Earth, in our physical bodies?

– There is a significant difference. Here's how: When we angels travel, sail, and float between time and space, we use gravitation. Otherwise, we wouldn't be able to hover. We can start and land through the power of thought, something you can't do because you need gravitation to move and not float away. Thus, there's a clear difference in how we move forward, but none of us would survive without gravitation. Without it, concepts like time and space wouldn't exist for either of us.

– How can we learn more about gravitation in the future? I assume we humans have only scratched the surface of this knowledge.

– That's correct. Everything you have learned about gravitation so far is only in its foundational phase. There is much more for humans to study, which you will do, but you are not there yet.

Gravitation is actually much stronger and more attractive than what is written in some books. Therefore, you will revise some books, with the authors themselves expanding their concepts and contents. After that, you will discover other and better applications.

– Can you tell us something that humanity needs to research further?

– Yes, we are now sharing information about gravitation and its function because we want to address human questions, including how gravitation governs and affects its surroundings (we will discuss this in more detail later). But before we get there, we must first study how the small gravitation works – soul development. This is where you will gain more knowledge and new insights into gravitation, as you have also been misled and learned in the wrong order.

THE SMALL GRAVITATION

Archangel Metatron speaks:

When people discuss gravity, they usually talk about the big one. Why is that, we wonder? Because it's always in the small gravity, the one we have within us, the one that holds the soul together, that we also find the answers to the big questions. Therefore, both now and in the future, we will illuminate our own gravitational movement, because if we learn and understand the strength that we all have within us, we can also use that information to change our own way of living.

We mean: If the soul were not created by something that holds everything together—our own small gravitational field, the one that surrounds our aura – we would neither be able to live on Earth nor in the spirit world. That's why it's so important that we first study our own small gravitational field before we can understand other larger driving concepts. Otherwise, we won't progress.

– What does our small gravitational field look like, the one we have within us? I ask Metatron.

– Our own small gravitational field consists of your entire inner self. But it's an outer field that binds everything together and influences a movement, just like your circulatory system and so on. Without our own small gravitational field, we would simply stop functioning, as would the universe. That's why you miss out on a lot of valuable knowledge about gravity.

So if you only focus on the big, because you think the big is more powerful, that's the only thing you'll reach. Without the small and the big of the universe, and without the cooperation between them, we cannot live. Everything is connected – small as well as big.

– Don't scientists already know that?

– To some extent, but there are also many scientists who are not at all interested in the spiritual body, so they also miss many discoveries that exist in the universe.

Think of it this way: Everything that we and you are created from comes from the universe. If the universe has a gravitational field, then you also have a gravitational field. If the universe has an atom, then you also have an atom, and so on. Everything that exists in the universe also exists in your physical and spiritual body. So to learn more about the universe, you first need to learn more about yourself. If you do that, you will also find big, significant answers to the questions you often ask us. That's why we need to open up our own inner self to open up the inner universe.

Understanding how the soul is created
is the study of the universe.

If we don't want to believe in the soul or the small, then we can't approach any great future answers either. Our entire development will come to a halt. Therefore, a more spiritual elite will now take over even in this area of research. They will then highlight how important it is that we first understand the soul before we can understand the universe. That's why we are now working a bit more intensively with some lightworkers so that they can spread their knowledge further. But before they can do that, they first need to reach the small within their own inner self.

Today, many people seek an *external* explanation caused by an *external* event. In the future, we will seek an *internal* explanation for an *internal* phenomenon. That is how we will increase our knowledge in the future. It seems that, even though we understand that the universe may appear large on the surface, we would rather strive for the grandiose than the small, the abstract, because we believe that is where the great power lies. Therefore, and to understand future research, we first need to reach the study of our own body before we can reach the study of the body of the universe.

The difference between current and future research is therefore based on your inner energy and understanding, and this is what many of today's researchers miss or are not interested in. That's why we need a more spiritual elite that can delve into the small—into your inner soul.

To help humanity see better,
you create more and more telescopes.
But in the future, we will first want to understand
before we learn to see.

The Quantum Theoretical Doctrine

We shall use what we call *the quantum theoretical doctrine* to explain further. It is here that you will develop the doctrine of the small gravitational field that our body and soul consist of. When you understand this, you will develop knowledge of *all the elements in the small gravity.*

This is also where some of your teachings have become misleading because there are those who do not want you to reach all parts of the human body. Therefore, there will be some confusion at first when new fundamental concepts within this branch of mathematics are renewed. Thus, the quantum theoretical doctrine will gain a new teaching in the future.

A book on this subject, with the same title as above, will also be published, where you will learn about new fundamental concepts that both our body, soul, and the entire universe consist of. And this, my friends, is what we need. The reason is that not many highlight the strength of small gravity today, at least not as we do in the spirit world. And since humans often seek knowledge of the big, they sometimes exclude the small, because, according to humans, everything small belongs to the quantum theoretical doctrine (particles smaller than an atom).

Why is that, we wonder? Even the smallest force can be worth its weight in gold and can be represented in its mathematical calculation. It is only humans themselves who stop their progress in the small because they stick to the old calculation system.

According to our source, the one we have in the spirit world, the small can be calculated in the same way as the big, and it is also possible to combine them. Therefore, we believe it is possible to apply Einstein's theory of relativity even within the quantum theoretical doctrine. How? Well, to understand the universe, it is always important that we first understand that everything is included in everything, that we can combine everything. So why wouldn't we be able to refer to Einstein's theory of relativity within quantum physics – within the small?

Why do you limit yourselves? Everything humans do is categorized, including the way they calculate, and then when something goes wrong, your whole train comes to a halt. Therefore, we must broaden our horizons and trust what already exists, what already plays its part, without us adding a mathematical calculation.

Sure, we find the small in particles; that's not what we mean. But what humans have missed is one final calculation within the branch that Einstein belonged to. When you do that, you will also understand how we can link the small with the big. But don't worry, the calculations will take place within a few years. It is also here that the next generation will take over and highlight this void, this content, that despite everything, nothing is empty. That everything can be calculated, big as small, empty as full. It is humans themselves who hinder their own progress.

Therefore, there will now be a major development to convince those who doubt that everything can be measured or combined. We will now briefly tell you who will further develop this in the future.

Ulrika: God's spiritual guide now enters my channel. He is the one who will tell us who this wise man is, and I write down exactly as it comes and without correcting his words.

Thank you, Ulrika. I shall explain:

He is a prominent man in the jungle where he works. A Frenchman who has come to stay and to further develop this part and take over the throne that has long sealed our physical instruments. His progress will highlight all aspects of the universe and how everything is connected between time and space.

He is unique in his character and can sometimes appear lazy, but he is not. He is just very much in his head and may doze off now and then, as all the inner activity takes much of his energy. But he is a good man, special in his kind. For he works for all to watch over and act in likewise.

He has an underlying temperament that can be misunderstood by many, but he is not temperamental. He is just unique in his mind and always has many irons in the fire. He is good, so do not misunderstand his disposition. And he is needed for future days, so give him time to develop. He is skilled. His inner knowledge may even earn him a prize in his old age, as it is only in old age that you will understand him.

He is gentle in his demeanor and is kind in his way. A man with a beard, rough in his features but soft on the inside. A Frenchman who always wears his hat at an angle.

He is a man who lives by the old school in today's times. So give him a helping hand. Lift him up for the fantastic work he does and follow in his footsteps. He will leave his mark that you can follow. So continue where his work ends and further develop that part just as you do with Einstein's.

It is also we who have assigned him this task because everything we do should be available to everyone. And he is a man who shares everything, including opinions, which confirms his value of gold. He has now attended our school and will therefore convey this fine and so important knowledge to you in the future.

The reason we are writing about him now is that when his greatness prevails, in old age, Ulrika will no longer be on Earth.

When we learn and understand the small, we will also expand the knowledge of the big.

THE EXTENDED ARM OF GRAVITY

Archangel Metatron speaks:

It is of great importance that we have gravity on Earth; otherwise, you would float, and the apple wouldn't be able to fall to the ground. But gravity can also control itself through motion, and it is this aspect we want to address when we talk about the gravity of the universe. Therefore, we can never control it or try to steer it in different directions. Thus, we can never make claims on it, which you will soon discover.

Gravity is free to move forward and at whatever speed is present there. And it's not really surprising that this is the case – that time and pace occur differently in the universe, the spirit world, and on Earth. Otherwise, we would not be able to operate between time and space or exist in the spaces of the future and the past. That is why we call it the extended arm of gravity, for it is gravity that extends time between spaces.

– How is it that time is so different? I ask Metatron.

– Gravity is everywhere, but what humans do not experience is the free movement that exists there. This means that everything is free to move. But at the same time, everything also follows the lines that are laid out both for you and in the universe. That is why it differs. This is also where we come to time since it is the time interval between spaces that differs. That is why time there and time with you are so different.

– How can the time between spaces affect us whether we feel gravity or not?

– Everything is determined by a rhythm, but that rhythm does not need to develop or open up in the same way, in the same place.

We will briefly take time and space as an example (We will delve deeper into time and space later):

If time in the universe were to stand completely still, no change would occur. But in the universe, we are not affected by time in the same way that humans are. But if space were to move, time would also need to do so; otherwise, we would not move forward and no movement would occur. And since the universe is not affected by time in the same way that humans are, it means that you feel weightless when time stands still. It is the free gravity that gives you a different kind of flight experience than you get on Earth.

– Then one might wonder why birds do not fall straight down to the ground if gravity is so strong?

– Birds can fly because of their wings. But the reason birds do not fall straight to the ground is also because of the universe's magnetic field – magnetism. So, if our magnetic field and gravity did not cooperate, the birds would not be able to fly. On Earth, we are pulled down by gravity, but that is not the case for birds. Birds actually fly in a sort of middle sphere because that is where all the cooperation between gravity and the magnetic field takes place.

– Why aren't we affected by this so that we can fly?

– You have no wings, and if you did and were constructed in the same way, you would also be in this intermediate state.

The Universe's Gravity: Connects.

Earth's Gravity: The law of gravity keeps everything in place so the apple can fall.

We can also mention that some of the mathematical calculations you have made concerning gravity are not entirely correct. They are not correctly calculated. The part that needs updating, and which we mentioned earlier, is that of the surroundings. Gravity is always owned by a surrounding concept, surrounding particles, that exist and are active within the same field. But the particles cannot act without their own gravitational ability, because both need each other.

– Do the particles need gravity to move, or do they have their own? I don't understand.

– All particles can act on their own, and this is something that some scientists have not quite understood yet, not entirely anyway. We also need our own gravity, but since we move by the power of thought, it acts a little differently than it does with you.

We know that this may sound complicated, so we will not delve into this now. But we will develop our discussion in the future, especially when Einstein's theory of relativity is further developed by our fantastic scientists.

GRAVITATION AND MAGNETIC FIELDS

Archangel Metatron speaks:

The part we want to convey now is the force of attraction that exists in outer space. We also want to explain how you can apply and connect it with your magnetic ability, in the same way that the universe does.

We will also discuss how a magnetic field can stop gravitational waves from the universe. This will be crucial for you in the future when you need to set everything in motion. Specifically, how you can apply this forward-moving technology even in your vehicles. You need to understand the ability of gravity and how it works in harmony with magnetism, as it is also a unifier – it brings everything together. But it can also, like a magnetic field, separate everything. That's why they work so well together.

However, you cannot adopt Earth's gravitational technique because it acts more as a force of gravity than as a unifier. It's designed to keep everything in place. So our advice is always to see how the universe operates, not how humans do.

– How can we adopt a gravitational movement while using a magnetic field that stops it? I ask Metatron.

Let's take a car as an example:

When you generate a gravitational wave in a car, which you will do in the future, you can keep it going along with the magnetic movement. In that scenario, it doesn't act as a stopper but continues as long as there's a gravitational range left.

It's only in areas where the gravitational wave is so strong, and where there is no magnetic field to counteract it, that it could cause a negative event. We do not want this. That's why it's so important that they cooperate. If you achieve good cooperation between both, you can also make the car just keep going, just like in the universe (we will continue to discuss magnetism in part three).

Where gravity steps In,
magnetism takes over.

This is also why gravity is seen as an extractor, because it's through it that the gap increases, and this is where the magnetic field comes in. A magnetic field can go in and stop gravity so that the force does not increase too much. If it couldn't do this, neither we nor you would exist, as gravity would take over the entire universe. In that context, gravity is a dangerous game to play.

Magnetic fields also ensure that gravity doesn't become too weak. That's why they work so well together, as they control each other's strengths and weaknesses.

So without the constant regulation of a magnetic field, the gravitational force would create more collisions than already occur. This is what gives the magnetic field a more decisive character, as it also serves as a balancer between the weak and the strong. So without a magnetic field with these important functions, everything in the universe would be a bit chaotic. Therefore, one can never completely eliminate the other. That's why there is always a good balance in the universe.

We know that humans have already discussed both gravity and magnetic fields for many years. Therefore, it's also high time for you to further develop this knowledge, or we won't progress. That's why you need to explore new perspectives on gravity.

Our advice is: draw lines where you normally wouldn't draw them. Connect matter that normally wouldn't be able to connect, and take advantage of the universe's magnetism. Try to utilize the universe's way of observing and controlling. Then pass on your knowledge to others, because it's only in this way that we can achieve success and create a new and better world.

The inner force that a magnetic field represents is also the one that will redraw your map in the future.

EARTH'S MAGNETIC FIELD

Archangel Metatron speaks:

For a magnetic field to function both as a stopper and as protection, it sometimes needs to be strengthened because any force used for protection can also weaken a magnetic field. And since Earth's magnetic field also acts as a stopper, it needs to be strong and stable, which it currently is not. That's why Earth's magnetic field is in the process of both strengthening its force and changing its formation.

A magnetic field consists of *small electro-spherical particles*, as we call them. These are also the unifiers we mentioned earlier. But sometimes a magnetic field can break, and that's when it needs to be reinforced.

– How does it do that? I ask Metatron.

– It happens through heat from other components, including the sun. When a magnetic field connects heat with other particles, their numbers increase, and that's how it repairs the weakened link – the one that broke. It's through a large number of highly charged particles that we can create, repair, add to, and change both our outer and inner world. This is also how we can add more magnetism to Earth.

– How is the heat created?

– Heat is created by the expansion of particles. The universe is filled with particles at rest that can be used when needed. When new particles come in, they bring a new force that together generates new energy, which in turn creates heat. This means that more magnetic particles are added. We won't go deeper into this right now, but we'll revisit it later if necessary.

– Thank you, dear ones. I understand. But what causes a magnetic field to break?

– It usually happens because of gravity. Think of it like a rope. The more you pull, and your opponent on the other side does the same, the heavier it gets for the middle of the rope. That's why it eventually breaks.

– Can we repair the soul in the same way? I mean, through particles that generate heat.

– We angels (souls) use a different method. It's a more healing and restorative method that some of our earthbound souls already use. It's a method that adds more magnetism, and by doing so, we can also increase the soul's energy absorption capacity.

This is what it's about when we talk about tiny particles that we then use for ourselves, for example, in our aura field. It's through magnetism and Earth's magnetic field that we can change these tiny particles and adjust how we want to transform them. This is also what's happening now with Earth's magnetic field, as it needs to change its function.

– Will this affect other parts of the universe as well? I'm thinking about Mother Earth, who is also healing and improving her inner workings.

– Everything in the universe has either a controlling or an additive function, where the controlling part has a more decisive character. It ensures that too much harm doesn't occur and that some parts don't take over others. The additive part ensures cooperation. In this way, they add to and control the entire universe. This is how we achieve balance.

If we didn't have these regulations, other parts could dominate and take over as much as they wanted. But to control and add, they need permission, and only if it's done in cooperation for the greater good. Therefore, it won't affect other parts of the universe, but it could have if we didn't have these regulations. This is what allows Earth's magnetic field to heal and move in peace, without risking being consumed by other parts.

– Who gives them permission to control and add?

– God does.

DISPLACEMENT OF EARTH'S MAGNETIC FIELD

The angels speak:

Earth's magnetic field has begun a shift. This shift is part of the new formation that will take place within a few years. Therefore, the magnetic field needs to strengthen its force. What we mean is that it will change its formation in a few years, which is necessary to reinforce Earth's field, both as protection against extraterrestrial intrusion, solar winds, and the solar flares that will significantly increase in the future.

– Will it affect us humans? I ask the angels.

– To some extent, but like everything else, it primarily affects our hypersensitive souls when we touch on topics like these. This is especially true for those who follow the teachings of the universe, just as you do, Ulrika.

– Doesn't everyone?

– We all do, but not everyone comes from the same place or the same original star. Therefore, you also react differently depending on what your soul is accustomed to there. So, some hypersensitive souls, just like you ... maybe you want to share how you have and continue to experience this time of change. Even though you didn't know back then that it was and still is our magnetic field affecting you so powerfully.

– Thank you ... I'll try.

It's as if both my inner and outer energy levels have risen, and there's no way to turn it off. There's no stop button. It feels like an overcharged battery, if such a thing exists, and I can't concentrate.

Imagine how it feels after a quick run, how the body continues to stay in that charged state even after you've come home and before your pulse has calmed down. But in my case, it's not my pulse that's high, it's my inner energy level that's elevated. It's hard to explain, but I simply get hyper. And since I'm someone who hears what the angels say, that's also where the power lies, which increases during times like these. That's why my thoughts sometimes spin out of control, and there's no way to stop them.

I've also had issues with hot flashes and reactions to some supplements that I don't normally react to, including B and D vitamins. It's especially vitamin B that's been difficult for me because it further increases my inner energy level. So, if I take supplements or eat food that gives extra energy while the sun, moon, and our magnetic field are affecting me, it doubles my inner strength, which becomes more than my physical body can handle at the time.

That's how it's been for me, and it still is. But I now understand that this has more to do with my sensitivity to the universe because I follow its teachings. I also understand why these days, weeks, months, and years have created so much frustration within me and how comforting it is to finally know why. Will it calm down?

– Yes, your energy level will soon adapt to the new. For it's not just Earth that needs to change and repair its magnetic field. Throughout time, as Earth and the universe have changed, so too has humanity, because we follow the rhythm of the universe. Otherwise, it would be difficult to exist on Earth.

– What happens to us humans when Earth's magnetic field changes its interior?

– What's happening right now is that the energy is entering the cellular level of your bodies, where it applies the same technique and the same energy that part of Earth's magnetic field consists of. And remember, even our smallest components have a surrounding magnetic field as protection to hold everything together. That's why it's so important that we first understand the small.

This also means an application for you, where your inner self will change at the same time as the universe changes its inner self. Otherwise, we cannot develop or manage to meet a new society where we live more freely and in harmony. So, to partake in this entire upgrade, you sometimes need to be in a physical body because that type of integration cannot be carried out in the spirit world.

– What happens to those who aren't living on Earth when this integration occurs?

– They too will reach this integration, though on a somewhat smaller scale since parts of it happen here in the spirit world. Therefore, the soul needs to be in a physical body to receive everything in this upgrade. So those who aren't living on Earth now will experience it in another life or on another planet, where these upgrades have already taken place.

The angels conclude:

Much of the future's research will take place now, which will lead to a greater understanding of the magnetic world, which is important. Because if you understand the universe's magnetic force, including what's happening between Earth, the sun, and the moon, you will also understand how stars and magnetic fields can guide you to a better life.

Some researchers have already begun exploring the large magnetic field, but they don't yet fully understand the force or how to create that type of force. But it's a journey you will undertake in the future, where you will both learn and use the universe's magnetic force and understand where it comes from.

This is where the Kathremes come into the picture. That's why they now want to briefly share how you can increase your knowledge of gravity and magnetic fields because this will lead you to the magnetism of the future.

THE MAGNETIC SPHERE

Kazandra speaks:

The ability we use on our planet is what we call *the magnetic sphere*. It is a foundational technology that you will also use and further develop in the future (the magnetic sphere is just an introduction to what we will discuss in part two). But before we begin, let us say this: the technology we use works for us because our atmosphere and its associated magnetic field are a bit different from yours. Our planet has an enormous inner core, which allows it to maintain two magnetic fields.

Namely:

1. We have a large magnetic field that surrounds all of Thrinica and is in direct contact with us. This is what we use to integrate with one another, and through it, we propel ourselves forward with a magnetic force.

2. We also have an outer magnetic field, which acts as protection against the powerful gravitational fields that sometimes affect our world.

Enough about that; we just wanted to highlight that there is indeed a clear difference between our worlds. That's why we work where you are, not where we are, so you can operate based on the power you have on your planet. For it is the power of the universe that you need to study now, including the small force surrounding your planet. If you study it, just as we have, you can discover a new form of magnetism combined with gravity.

How do we reach the magnetic sphere? To explain this, we've divided it into three parts:

1. First, we need to get close to a gravitational field, and to do that, we must first lower it, which we do with the help of a strong magnetic field. When we do this, we establish a sustaining balance.

Once you've found that and the difference between time and space, you can apply the same technique on your planet. This will positively impact your results if you first learn how gravitational ability and magnetic fields differ, but also how they work differently in various dimensions and across all times and spaces.

So, start by understanding this difference, and then try to find what we call *close cooperation* in different concepts. If you do this, you will better understand the foundation that you need to grasp first before you can create a new movement scheme, supported by this foundational technique (We will discuss the movement scheme later).

2. Once you've reached that point, together with our world, you should learn about electromagnetism and the powerful energy emitted by a neutron star, especially the part where it transforms its outer layer to increase its internal energy absorption. Part two is thus about the study of neutron stars and electromagnetism (We will discuss neutron stars and electromagnetism later).

3. When that part is complete, you need to study free radicals more than you currently do. Free radicals are also what we use on our planet, and they are what open the door to close cooperation between gravity and magnetism, as well as other elements we mentioned earlier. But the most important thing we advocate for, and what we want you to use, is helium, which we refer to as carbon. You will understand why when you learn what helium consists of and where it originates from.

This is the foundation for all teachings within the magnetic sphere. But before you can comprehend your new, improved electromagnetism and the movement scheme you will create in the future, we must first explore the realm of stars. It is there that all transformation occurs. It is there that you will gather the power you need to set everything in motion and that works in harmony with gravity and magnetic fields. Everything is interconnected.

CHAPTER 7

STARS

ANCIENT SEAT OF LEARNING

Angel Sepatzon speaks:

We have now finally arrived at the stars. This is where everything really took off, where everything exploded into its full effect, where the foundation for creating our first star was laid. Despite this, we will not delve into how a star is created; you already know that. What interests us is how we can adopt their properties in our new power plant. The purpose of this section is to teach you the technology you will use in the future, based on the neutron star's way of harnessing and acting.

Therefore, it was no coincidence that humanity discovered the neutron star many years ago, as it was part of the plan we have in the spiritual world for you to use this technology in your future fuel systems.

But before we continue, you first need to understand what you have lost. That is why we will start with some discoveries made by the Romans, as they were already onto what we know today. It was a time when humanity looked inward to find external answers, which they did by studying stars in a more introspective way. This is what we want you to return to now: finding the inner capability they used before you can understand and create a new power plant.

– Why do we need to go back in time? I ask Sepatzon.

– Because it was after that time that humanity lost its inner capability. There is actually a lot of prehistoric significance in this that we believe humanity missed but the Romans did not. The Romans didn't just see a star; they also understood that without stars, it would be dark. They didn't know what a lamp was, but they understood that a certain amount of light was needed for a star to shine. But since there was no electricity or light bulbs at that time, they did not grasp the development of electricity. What they found instead was the power-carrying chain of their time – a candle.

They learned that without oxygen, a light could not burn. Believe it or not, but even back then, they had a good understanding of how a light is affected by the external environment. They also did everything in their power to further develop the technology of light, for example, by enclosing it. They believed that without the influence of wind and sun, light could not burn. The idea was to better understand the external and internal construction of the candle and how it is affected by the external, as that was the only way they could create sustainability, said these men.

To you, this might sound like a ridiculous story by today's standards. But think about how significant it was at the time to try to understand the existence of the universe.

Furthermore, they also learned to master water as an energy source, and by the standards of that time, they would surpass today's researchers by leaps and bounds. It was a technology we used on Atlantis but had to shut down, and we wish for you to revive it. This was Ulrika's role in one of her previous lives, where she worked with the different structures of water and as a leader of the water energy we had at that time, and which we will now bring forth and use. But until then, we will stick to the technology the Romans used.

– What does this have to do with today's stars?

– Actually, the Romans got their most brilliant ideas from the heavens, including water transportation. They had already developed a well-thought-out method for transporting water, which unfortunately was lost over time. But it is precisely this method that we want to revive now, and to do that, we need to return to the technology they used – the inner understanding they had of Mother Earth and the universe.

So what did the Romans do to study and understand the universe? They mapped important geographical points to observe how a star expels its dust, as they called it. Through this, they could track the journey and movement of the dust, and all of this was done through their intuition. They also understood early on that the universe could not shine without a certain number of dust particles, which we call photons, which was a very good and early analysis.

The Romans also knew that when it came to water and light, they were onto something big. Therefore, they locked away their recorded knowledge so that only the scholars of the time could shed light on its strength and understanding. People were somewhat selfish at that time. They were, as mentioned in today's history books, the rulers of the time and wanted to remain so.

They simply forbade unauthorized individuals from entering

their realm, even the scholars of that time, which eventually negatively affected them as well because much of what they had learned was lost in the fire that occurred. Therefore, there are not many records left from that time, which is a shame.

But we have still managed to recover some, notably through hieroglyphs (hieroglyphs are pictures of animals or objects used to represent sounds or meanings. They resemble letters, but a single hieroglyph can represent a syllable or a concept). It is thanks to them that we can now look back in time to learn and see how they did things back then.

The Romans, therefore, already had a better understanding and sense of the calculations of their time than many do today. At that time, they did not have the technical equipment we have today, which meant they had to rely on their inner sense and what the universe provided them. However, much of this inner, feeling-based understanding no longer exists today. The more technology humanity develops, the less you sense inwardly. And the less you sense inwardly, the less you understand the universe because the universe exists within you – in the soul. This is what we believe today's humans have lost.

This is what we want you to take away from this story: that not everything needs to be learned through expensive telescopes. A good inner awareness and a knowledgeable teacher by your side who has a good understanding of the universe can develop human knowledge more than a technical aid ever can. Not everything succeeds the more technology you develop; some of it can even hinder your inner sense. That is why the Romans achieved much more than you do today, despite all the technical equipment.

Therefore, we will now enter the realm of the stars. It is there we find our teachings, and as mentioned, the Romans had already cracked that code. But since much of the information from that time no longer exists, it is up to you now to revive the research the Romans discovered back then. We will start with the fundamental purpose of the star.

Something extinguished the
soul's intuition, which also extinguished the
knowledge of our original star.

STARS

Angel Sepatzon speaks:

The creation of a star consists of three fundamental intentions:

1. A star *processes* the gas present in the universe and other material nutrients.

2. It acts as a *stabilizer* to enable the transformation of matter.

3. It also functions as a *brake* to ensure that the star explodes at the right pace and for its own sake.

For a star to form, a foundation is needed, and as with everything else, it is crucial that this foundation consists of stable material – in this case, gas. However, for a star to grow, it needs to supplement itself with additional material, which it acquires from the universe's dust, including photons. These photons form a strong, solid light that makes the star shine.

The star also serves as a stabilizer to prevent the gas or other nutrients from dispersing or floating away. We're not talking about gravity pulling matter together; we're referring to the internal process that keeps the core stable and strong. This is where the transformation of lighter and heavier elements occurs.

To prevent a star from exploding prematurely, it can also act as a brake by pulling inward. This is what it does before it explodes. It contracts into a small hole before releasing all its internal power in an explosion. Other parts then capture these nutrients, and in this way the process continues.

– Why does it act as a brake if it's meant to explode? I ask Sepatzon.

– It acts as a defense mechanism to avoid its own destruction, which it does not consciously know but is simply a reaction when it starts to feel full. Without this brake, there would be even more star explosions than what currently occurs in the universe.

Now, you might understand how crucial the universe's regulations are. Without them, stars could act on their own free flow and not be constrained by the universe's rules. Moreover, if all stars released their internal power simultaneously, it would negatively impact the universe by disrupting its balance.

Balance between heavy and light.
Balance between old and new.

Balance between death and birth.
Balance between black and white, and so on.

That's why our stars have an inherent brake, to prevent things from moving too quickly or releasing too much matter at once. What do you think would happen if it moved too quickly? The star wouldn't have time to fully process its internal matter and undergo the transformation before it explodes. We would also lose a significant portion of the star's essential elements, especially those that take a long time to develop.

A star can never act on its own because it is filled with matter that affects its power, and this is what we will use in the future. But we need to learn more about the star's contents. Much of the information a star holds has not yet fully settled within you. Therefore, there are still some puzzle pieces left to study, such as what happens before an explosion occurs and why it doesn't just stop. While it is the star's internal power that causes the explosion, you need to investigate whether it receives assistance from other external parts of the universe, which will further expand your knowledge.

We cannot and should not reveal more than that, as it is up to humanity to find the answers to these questions. What we can say is that what a star can do is a so-called restart.

When a star begins to feel full, after producing most of the lighter elements, it can process its contents again. This is how it undergoes a restart. A restart means it begins where it left off and can then stay in the same stage a bit longer than usual. This is how it extends its own life.

– Doesn't this affect the star's contents – the substances the star produces?

– No, all substances remain intact because the star stays where it last stopped.

– Why does it restart?

– Through a restart, the star gets a chance to rest. This also prevents too many stars from exploding simultaneously. This is how we create balance in the universe by distributing, resting, producing, and repairing, and so on. Everything in the universe always happens at a moderate pace.

– Doesn't the star become cold when it enters rest mode?

– No, the star never becomes cold inside because there is always active movement occurring.

Let's use a car as an example:

Imagine a car idling – it is running even though it is completely still. In contrast, a car that is moving forward is in motion. The idling car remains warm because it is still operating and using fuel, though not as much as the moving car. The moving car will wear out much faster than the idling car because it will either eventually break down or stop when the fuel runs out. Other cars, other stars, then take over as producers while others rest. This way, they maintain balance.

– Can we observe anything when a star rest?

– No, everything happens internally, which is why we chose to include it in our book. Even the Sun can undergo a restart, reviewing its internal state and thereby extending its lifespan. This will also happen now until we replace our current Sun.

It is a restart that you will also use in the future, making it very important. But to do that, you first need to understand the star's three intentions. You will use this transformative method in your new power plants. But before you can do that, you must first understand the power of a neutron star, as it will generate the energy your new power plants will distribute to your households (we will discuss your new power plant in more detail later). But before we continue to the world of neutron stars, let's conclude with a brief note that Ulrika saw during an astral journey.

Ulrika speaks:

I am standing in front of a brilliantly shining star. There are many of them, and they are much whiter and shine much brighter than we are aware of today. But that will come when we reach the part of the universe I was just in. It was a journey that took me beyond our own universe. It was like a completely different world, a whole other universe.

I also saw many *star planets*, as the angels call them, and what they do cannot be explained now, but you will learn about it in the future. I see the star planets as very small, and I get a sense of emptiness I have never felt before, like a completely new foreign world, but there is no life there, not like it is among us.

NEUTRON STAR

Angel Sepatzon speaks:

A neutron star is what remains after a star has exploded – what's left after a supernova. As the name suggests, it consists of both neutrons and other components scattered from the supernova itself.

Just as Earth attracts the energy waves of the Moon and Sun, linking them together, a neutron star does the same. However, the neutron star goes a step further by deliberately collecting surrounding external material to continue growing. The more material it gathers, the faster the neutron star spins. The magnetism and controlling properties inside a neutron star are what humanity will use in the future.

By harnessing this controlling and guiding force, you can achieve a better environmentally-friendly fuel. But to apply the power of a neutron star to your future vehicles, you first need to create something that can operate in a similar way. What a neutron star does is essentially twist itself apart and back together, which is its primary function to create new material. Thus, our neutron star also needs to twist apart and back together to function as desired within our technological applications.

How does it do this? It spins, and the enormous power generated by a neutron star can be used as a driving force in the future. The spinning of the neutron star, or its rotation, is an eternal process where the magnetic motion never ceases, making it an excellent solution for our future technology. We will also use the beam emitted by the neutron star as an additional power source in our future vehicles.

– Why does it spin so fast? I ask Sepatzon.

– After a star has transformed into a neutron star, it is free to act according to its own power, which it does through spinning. It is through this spin that it gains more energy, which is necessary for it to expand its contents. Without these beams, the neutron star would explode. Therefore, it needs to release some energy to then gather more. Like everything else in the universe, it needs to achieve a balance.

This is also how it creates its incoming and outgoing ports. You can also create these ports on Earth, and when you do, you introduce a magnetism based on the properties of a neutron star.

– How can the neutron star's beam affect our future technology?

– We will use the neutron star's beam as an extension of the car's power, both as incoming and outgoing ports. Every time the spinning stops, the beams will restart everything. It will function as an extra energy source when the old one is depleted. Quasars also have these properties, both as intake and output, which we will discuss later.

– I still don't understand. Could you elaborate on your thoughts?

– We'd be happy to. We understand it's challenging

Consider a "light spinner yo-yo" as an example. The type that children play with, with a string attached (a light spinner yo-yo is a variant with built-in LEDs that blink when the yo-yo spins):

When the child pulls the string, the spinner starts and spins, creating a beautiful light show. The conditions are similar here, and this is what we advocate for with the neutron star's laser beams. They should function as a start button when energy begins to run low. The strings on each side represent the incoming and outgoing ports, providing this extra energy after the neutron star's actions.

Now, some might think: Surely, we can't mimic a toy in our advanced and technological society? Yes, we say. Why should the most advanced results always come from the most complex constructions? Some problems are solved in the simplest ways. That's how our ideas emerge – in simplicity.

Ulrika tells:

Through an astral journey I took with Archangel Metatron, I saw further how the neutron star is constructed. He showed me a timeline and explained that on the timeline we are currently on, you cannot see this because you have not yet developed to that point. But when we shift timelines, you will also see what he showed me.

He revealed that there is a time axis in the center of the neutron star, and this determines whether you can see that the neutron star can also rotate, which it does when it emits its signals, as the angels call them. We're not referring to the spin you see today when it spins upright. This is a spin that happens in all directions, even more than you see today. That's how I saw it – spinning in all directions simultaneously.

It's as if it can also snap, standing completely still in an instant. This *shift* is what the angels want us to develop – rapid rotation to complete stillness in a few seconds. Depending on which timeline humanity is on, you may or may not see this, and to do so, you need a cascade telescope.

Fredzo said: "Try to observe how it behaves, Ulrika, not just how it looks from the outside. We will assist you with this. It is a significant lesson already present on our planet."

Archangel Metatron said: "The line you are on, Ulrika, is slightly ahead of its time, which is why you can see this but others cannot. That's why we have opened that part to you now."

I stand before the neutron star and feel a profound sense of wonder. You know that feeling when it tingles throughout your soul, when all positivity spreads through your body. And it was an inspiration greater than I could ever have imagined. It was then that I realized the neutron star is for me what black holes are for Stephen Hawking. I have finally found my universal calling! My soul experienced such immense joy that day.

Summary:

By learning to study the behavior of neutron stars, how their power and motion increase as they gather more external material, we can increase the speed of vehicles. This is how we can improve and control the fuel in our cars – by harnessing the power of the universe.

The light beams emitted by the neutron star, *flashes* as we call them, symbolize the power the neutron star creates internally – energy must be released outward. Consider our incoming and outgoing ports; otherwise, it would eventually explode, especially at the immense speed inside and around a neutron star. Thus, the release of energy outward is fundamental, even when we adopt this thinking in our vehicles.

But before you fully embrace this new technology, you need to understand where the neutron star's fundamental power comes from. For this, my beloved friends, you can only observe in the ancient part of the universe, where time began and where everything subsequently developed. Archangel Michael wants to join us and explain.

Archangel Michael speaks:

In reality, the power of the neutron star does not come from the source that humans believe it does. According to the knowledge in our library, in the spirit world, the fundamental power actually originates from ancient times – a sacred force created long before our first star and is what our stars reach.

This is the very foundation of a neutron star's power, which we refer to as when something collected from ancient times has further developed in the new era. It existed long before our new universe was created. Therefore, you too will reach this ancient source of information, but not until you have begun the concepts of new technology.

– How do we know this? I ask Michael.

– You will only arrive there as humans always believe it to be a coincidence when you have learned how to harness the power of a neutron star within a magnetic field. Then you will *solve* some newly discovered mysteries – not find them, but solve them like a puzzle that needs an answer. It is then that you will discover the ancient part of the universe, more than you have already. So seek, look, and solve the answers to your future questions, and you will eventually find this ancient power. It is a fundamental source that only our old universe could create but is now increasingly emerging in yours.

Therefore, we will now revive the old era of our time's history, including our ancient power. For it is through forces like these that we can reclaim our kingdom – the time when we lived on Atlantis.

NEUTRINOS

Angel Sepatzon speaks:

To achieve the full potential of a neutron star, we need neutrinos. It is the spinning action that functions as a perpetual motion machine, generating the power and reaching the speed required for your new power systems. To understand what neutrinos are and how they work, we will now delve into the world of neutrinos – a world where everything spins faster than a hot press release.

Neutrinos are particles that serve as the energy source for neutron stars, the sun, and the entire universe. They also provide the energy that a neutron star needs to increase its capacity. The particles, or energies, are created and shaped inside the star, with the purpose of expanding. They can multiply faster than anything else in the universe – at least from a human perspective. So, the more particles there are, the faster the strength increases.

As mentioned, neutrinos are a source of energy. Almost everything in our book related to environmental improvement methods depends on harnessing more of the universe's energy, and it is neutrinos that will lead humanity to new, better technology.

– Why neutrinos specifically, and aren't we already working with neutrinos? I ask Sepatzon.

– Good question. It's true that many researchers are already using neutrinos very effectively. But there is still much to learn, especially regarding their scope. The surrounding aspects will become even more crucial in the future to better understand how they can multiply so rapidly and thus increase in number, which is what we want. This will be crucial for our future technology – having particles that can generate energy and expand quickly. When we achieve that, we can apply the same technology in our future machines. The machines we are talking about will need to produce their own power internally.

Neutrinos can also intersect with each other, which allows them to gain more energy. The more intersections we create, the more energy we provide them, accelerating the conversion process.

– How can we apply neutrinos so they interact with magnetic forces and create rapid motion in our vehicles?

– We do this by letting them act as drivers. They will determine the speed of the conversion process themselves. Thus, they will take on a leading role and will be responsible for the entire operation.

Therefore, it is now the task of researchers to create this high and rapid capacity, which we know is behind the expansion of neutrinos. And don't hesitate; neutrinos are actually much faster than you think. The speed of neutrinos has not yet been fully understood by your professors, which is why you hesitate. Even here, like with magnetism, the concept has been incorrectly presented to prevent full understanding.

- What is the true power?

- The power of neutrinos is much stronger than you realize. They can interact with many more factors than you understand. That's why you are uncertain about their real potential. Therefore, we recommend experimenting and testing them, even with what you believe they cannot be combined with. Try to add and expand their power beyond what is currently known or written about their development. Neutrinos can be expanded and developed more than is currently documented.

So, don't listen to those who claim they are weak and have limitations. This is incorrect! Neutrinos never limit themselves, and they can be added to the electromagnetic concept as well. They can achieve enormous speeds, and their power can become almost as strong as today's gasoline-powered cars, using just *a single* neutrino power source. So continue your research, develop them, but above all, expand them.

In the future, we won't need to refuel (though we will always need to replace and repair parts). The process will automatically replenish and store itself. For this process to be possible, we need energy particles that can replicate themselves at the fast pace they currently do. We cannot use slowly produced energy particles. If we do, production won't advance inside the converter at the desired rate.

Neutrinos also love to pick up hitchhikers along the way. They are incredibly coordinated, so feel free to combine them, which will increase their power. You can also line them up, and you'll see they always follow each other. This is what makes neutrinos so special – they always move forward, never backward. They can change direction, but only one way at a time.

Neutrinos sparkle and shine
in their full glory.
They can transform most things
within their class.

Neutrinos are indeed the concept of the future, and it is the core that will "elevate" (hint) future science. So harness these phenomenal particles, created from the strongest and most sustainable source, and use them as kinetic energy in your kitchen, home, vehicles, and as a heat-resistant component in your boilers. Our interest lies in this forward-moving kinetic moment.

We know that neutrinos are already a well-known field on Earth, which is why we haven't mentioned them more extensively. But you will understand what we mean because it is neutrinos that will create new solutions to current and future problems. However, we cannot create powerful energy without something to hold it all together, which is where the magnetar comes in.

MAGNETAR

Angel Sepatzon speaks:

When a star explodes and becomes a supernova, it transfers all its contents into the universe. What remains is a neutron star, known as a magnetar when it has an extremely strong magnetic field. A magnetar's magnetic field is thousands of times stronger than that of a typical neutron star. This means that the star's magnetic field persists and is far more powerful than any other magnetic field in the outer regions. This is the phenomenon we advocate for human use in future magnetic propulsion systems.

We understand that we cannot apply the full strength of the universe on Earth, as it would be too overwhelming. However, once you begin to use the same technology as a neutron star, including the magnetar, you will both see and understand how to harness the level of magnetism that you can manage.

To achieve stronger magnetism in the future, we recommend the magnetar and its method of operation. The surrounding magnetic movement is so strong that it can protect and set everything into motion. It is this movement that we are interested in with regard to the neutron star and its internal magnetism – using the magnetar's approach and external strength to unify everything.

The magnetar's properties have also protected many parts of the universe because its powerful magnetic field has managed to avoid many otherwise volatile problems. Without this protection, the compact material could cause significant damage in the universe.

For example, imagine a lead ball floating through space. This enormously heavy mass, generating heat, could burn and melt most things in its path, which must be prevented. This is why a magnetic field, especially that of a magnetar, is so crucial.

We are aware that the magnetar is already a well-studied area in science. However, our focus here is not on the magnetar itself but on its strong magnetic field, which we consider an important component for the future. Even though you may already understand the power emitted by a neutron star, including aspects like in- and out-portals and neutrinos, you will need a strong field to hold everything together. Only then can you embark on the next phase – a new power system.

Magnetic fields store energy.
This is the aspect we will utilize in the future.

CHAPTER 8

BLACK HOLES – QUASARS

STEPHEN HAWKING

Ulrika speaks:

Stephen Hawking is one of my spiritual teachers, and he assists me in the work we are doing together. He will also continue to share his knowledge through other spiritual channels who seek his guidance, as his future work will focus on the area he explored during his lifetime – black holes. "It's a tough nut to crack, but with a little help from above, we will soon get there," he says with a smile.

Therefore, we are now entering his realm, and it is a great honor for me to develop alongside Stephen and learn from our amazing and beloved scientist. He has so much love and knowledge to share with us. "By doing this, you will better understand why my journey was what it was," he explains.

I feel at ease with Stephen, and he always provides me with immense calm when we work together. So, thank you, Stephen, for your friendship and knowledge. You illuminate my soul with your incredible insights, and I know that we will continue our journey together and support the world even in the future.

Archangel Metatron speaks:

We will not only focus on the power of neutron stars in the future. Both black holes and quasars have long been well-known phenomena in science, becoming increasingly familiar as we have learned more about the fundamental concepts of stars. They have always existed; it was merely humanity that could not see them initially.

In this chapter, however, we will only discuss what we receive from God, our highest Lord, and the elements that Stephen Hawking himself selected and developed, though they remain foundational. Specifically, how can we continue to work with black holes and the quasar technology that humanity will develop in the future?

Therefore, this chapter is dedicated to his honor, as a way for all of us to pay tribute to him as a person and for the remarkable work he has always done. This is also a lesson that Ulrika is currently undertaking with us. As her sensitivity to all parts of the universe has increased, so has her inner knowledge.

BLACK HOLE

Archangel Metatron speaks:

A black hole forms when a massive star explodes. The collapse then leaves behind a black hole, which is governed by the gravitational force present throughout the universe. This force, along with the so-called "contente line," acts as a producer within a black hole.

However, for humanity to continue its studies and learn more about this incredibly mysterious black phenomenon, you first need to understand how it was created from the beginning. Understanding its creation is fundamental to understanding how it acts in its entirety and how it manages the material it absorbs, much like a vacuum cleaner. Although it may seem that way, that's not quite how it works. The mystery of a black hole is much more nuanced. A black hole actually began its journey when the universe was created. This is what we will delve into now.

Stars are not just here to produce; they are also here to transform the universe.

To create something, we need to add something that can then be combined. Without a black hole, we would lack a connector that can gather matter and then use it for its own power and survival. Otherwise, this cycle of creative power would not find its place. This was the basic idea, and it continued in that manner.

– Was a black hole created after The Big Bang? I ask Metatron.

– We can say this: without these collectors, the universe could not have created anything. Thus, the primordial force existed before The Big Bang, but in a different form. It expanded as more matter was created in the universe, which it could consume and grow from.

– I don't fully understand. How could there be a black hole before the first star was created?

– A black hole is part of a star where we first needed to create the force before we could create the actual product. The star's task was to create an internal force. What we mean is that the force was there beforehand, which then created a black hole as it grew.

Think about how a child comes into being. We first need to create an embryo; otherwise, the child has nowhere to grow. But before we can create an embryo, we need to create a force that can act within the embryo and ensure its growth. Thus, the embryo was not initially a fully functional embryo. Everything takes time to develop. Likewise, a child, just as a black hole is part of a star, is part of the embryo.

But the stars we are talking about were not in this universe. Therefore, their power was different and smaller in both size and strength. Thus, the stars only left behind a precursor to a black hole, and this is where the primordial force comes into play.

– So, the stars were so small that they only left behind a force and not a black hole. Is that what you mean?

– Yes, exactly. That was also the purpose of those stars – to develop the primordial force of a black hole (not all stars have the same purpose; each has its special role in why it was created). It was then the same force that began to gather matter. We needed to create the primordial force first, which was the very purpose of a black hole. And the larger and stronger the stars became, the larger and stronger their interior grew, and eventually, they left behind what we now call a black hole.

But a black hole is not only here to gather matter and grow larger. The force that God created for a black hole is also here to improve the universe, by cleaning away what could become, but what cannot become, because if it did, there would simply be too much of that kind. Or it might be created with a different purpose than what our universe can or can handle at the time.

– Was it this that allowed the universe to begin acting on its own and that a so-called "guardian" was needed at that time?

– Yes, exactly, and it's wonderful that you figured that out. Because that's exactly what they sometimes do – they clean, both to keep watch and to maintain order. Just like you do. Sometimes you also need to clean, otherwise, your house would be overwhelmed by dust or other things that do not benefit your world. You clean to achieve balance in your home, and so does a black hole.

This mysterious star thus has its own determining function and behavior that humanity does not yet fully understand, especially its interior. For it is within the interior that transformation occurs, and that is what humanity needs to learn more about, as it is relevant to the time we live in. Therefore, we will now, together with Stephen, highlight the lightning bolts, channels, and the contente line that exist within a black hole.

FLASHES

Ulrika speaks:

We will now take you on two of the astral journeys I took with Stephen Hawking. These journeys take us deeper into the mystique of black holes. It was also through these astral travels that I first met Stephen.

The first journey is about *the lightning bolts of a black hole,* and the second is about *the channel of a black hole.*

I sit down and breathe deeply and calmly. Suddenly, I find myself in a white bubble, floating up into the universe. Once I arrive, I open a door and step out onto a narrow white path. A man meets me, dressed in white at the moment. At first, I thought it was my ancient ancestor, my guardian angel, but later I learned that it was the archangel Metatron. We embrace and smile at each other as we meet again in the place where I learn in the spirit world.

I look down and see all the colors of the universe shining and sparkling in their clearest light. The color yellow is particularly prominent, but I don't know why.

I look up again, and Metatron shows me a balloon. He explains about the helium that humanity will use in the future. "When you have learned and adopted helium, more than you use in a balloon, helium will replace hydrogen. The hydrogen you use in your cars, which is not entirely safe. Therefore, you will soon switch," he says.

I look around. Further down the white path, a man is walking. A man in beige pants, a white shirt, with light hair and glasses, and I immediately recognize him as one of my teachers. He smiles as warmly as the sun itself, and we rejoice in our meeting. Who was it? I met my teacher, my wonderful, fantastic teacher, Stephen Hawking. He takes my hand and leads me up a staircase, and we stop at a large black darkness. It flashes brightly, and he asks me to observe.

– Try to understand why the lightning bolts exist, says Stephen.

– Are they the energy channels in a black hole? I ask.

– Yes, exactly, and researchers already know that they exist. But they don't function exactly as humans think they do.

– What do you mean by "not as humans think"?

– I mean that it's the energy channels that act as collectors, that the lightning bolts possess the fundamental strength of a black hole. That's why they can draw in so much from the universe.

– What are the lightning bolts made of?

– Humanity will have to figure that out on its own. I cannot reveal more than that.

– What creates the lightning bolts?

– A black hole is essentially a star, but just like a neutron star, it is created from an enormous amount of energy, otherwise, no lightning bolts would form. Think of the immense power of a neutron star and then imagine lightning bolts within it. That might help you understand how they can act the way they do.

– So, it's the lightning bolts that give a black hole its speed?

– Yes, exactly, and this is something that humanity needs to research further, as it is a process you need to go through before you can advance – to study lightning bolts and how they can fully manifest their strength.

On Earth, they emit energy, but in a black hole,
they consolidate energy.

I look directly into this gigantic black hole. I see how it flashes and sparkles, but he does not allow me to see more. "We'll continue tomorrow," he says with a smile.

I thank him for the wonderful meeting we had together. I conclude by thanking God for the beautiful fellowship we have in the spirit world. A place where everyone collaborates and helps each other so we can grow in body and soul.

I come back to myself and look back on this meeting as one of the clearest memories I have received from the angels. It was a memory that came to life; it was a memory I saw, which made it easier for me to go through. One day, at a place where we consulted with each other, was where I, together with Stephen, learned about the sacred power of the universe long before this earthly life began.

Eye to eye. Heart to heart, we met and could share
this beautiful moment together.

THE BLACK HOLE'S CHANNEL

Ulrika speaks:

The next day, we continue our journey. I sit down and breathe deeply and calmly, trying to capture the right feeling of what Stephen wants to show me this time.

Today, it happened quickly. I was immediately drawn into an enormous black space, and I felt alone. It felt desolate in a way, almost as if it was a definitive end. It was vast. Enormous! I saw lightning bolts again.

A tremendous force grabs me and pulls me into a tube, and the deeper I go, the faster it gets. My soul spins and spins, and it feels like I'm being pushed further and further in, but I don't know where. Suddenly … everything stops, and I start floating again. It feels calm and pleasant now because not much is happening here. Stephen gives me the sensation that I am inside a black hole, in its inner core. There are a few lightning bolts here and there, but otherwise, it's quite peaceful.

The color is gray.

Something grabs me again and propels me further. It feels like the outer part has started moving again. It's like when the force is pulled in; it's calm inside the core. But when the outer disk activates, when everything starts to be pulled outward again, the activity increases.

I'm ejected and wake up from my astral journey. A bit disoriented, but I feel relatively well and smile to myself because I know who was here showing me all this. I am so grateful.

We continue our conversation in the kitchen:

– Why is it relatively calm inside the core when everything is being pulled in, but increases when it is pulled out? I ask Stephen.

– It's because there is an activation during inflow, which is not as intense as during outflow, even though it might seem so. When the accretion disk activates in the outer region, it absorbs all external energy, and this external energy is actually much more powerful than when it reaches the core.

– Why is that?

– It is precisely in the transport phase that you experienced, where all the energy is collected. Once inside the core, the energy has diminished. When the energy then releases everything again, the energy increases again, even within the core, and that's when it emits its beams of light. But to do this, it first needs to build up an enormous internal energy within the core, which at that moment becomes much stronger than it was when everything was pulled in. There has been a transformation.

The increase in pressure is to avoid its own destruction; otherwise, it would explode or simply vanish. The energy is so great that without the rest that the core needs sometimes, the energy would become too much. That's why it sometimes needs to be in a resting state before everything starts again, and it continues this cycle until it disappears.

– Why does it disappear?

– Because it is finished, it has completed its purpose, and then new black holes take over. Otherwise, the entire universe would be governed by black holes, which is not supposed to happen. There would simply be too much material produced by a black hole, which it then releases, and we wouldn't achieve balance in the universe.

– I've heard that our early (old) universe was packed with matter and that's how our supermassive black holes formed. Is that true?

– Yes, but the universe didn't have the high density at that time that some believe. But that's not something we'll delve into here. I've noted your question, and perhaps we'll explore it in the future.

We continue with the channel:

A black hole also consists of a core house, as all stars always have a fundamental core from the beginning. Think about what we mentioned earlier – a black hole forms when a gigantic star explodes. After the collapse, it leaves behind a black hole. The core house thus constitutes an inner part of a black hole.

In the core house, all particles that a black hole pulls in are gathered and drawn into a stream-like line. There, they spin around like a long chain until either the energy diminishes or their ability increases. This means that some particles die, but only the weak ones. Some increase in strength, but only the strong ones. Others merge like a pair, and that's when a collision occurs. It is from that event that capacity increases and that's when lightning bolts occur.

It is inside the core house that all the power is created. It is also where it then transforms everything. After this transformation, it can send out its internal power again and then gather power again. It is also from here that it gets its suction capacity.

– How can the particles go in and out?

– They do so through a channel. There is indeed a channel that is not as affected by the pressure that occurs in a black hole, and it is where all the particles collect, those that go in and out.

– How can there be a channel in a black hole that is not affected by its enormous suction power?

– Because it is independent. It moves around the force but doesn't go all the way in. It is therefore not controlled by the rest of the black hole since it only acts as a mediator, so to speak. It transports – it does not act

Imagine a car during a storm, but inside it is completely calm. The car transports you (the driver) but it cannot act against the speed occurring outside.

– How do the particles end up in the channel?

– They end up there because they were carried there by the force when they were pulled in. Some stay in the core while others are transported further.

– Why do they do that?

– Because it creates discord, which is necessary for a black hole to neither stall nor rest. If it rests, no transformation occurs, and then the lightning bolts cannot expand.

– So, it's like a gas pedal in a car? If I press the pedal, the speed increases, but if I release the pedal a bit, the speed decreases. Can you say it that way?

– Yes, exactly, but in this case, they function as intake and exhaust handlers. As mentioned, it's a channel that transports, not acts, which it does so the core can continue its work of creating lightning bolts.

– Why does it create lightning bolts?

– Because they create electricity in the universe.

– How can a black hole disappear or die?

– A black hole never dies completely. It will always grow, but in another form, and this is something some researchers have missed – that they transform into something else. Only the strongest particles survive in a black hole. When a black hole only consists of weak particles, they fizzle out, but they never completely disappear.

It's somewhat like a soap bubble. It either grows larger and stronger or eventually bursts from too little force, and all its contents then fall to the ground. But since everything in a soap bubble is transparent, we cannot see it. And depending on how strong the surface is, the greater the chance the bubble will last a bit longer.

– So, it's the outer part of a black hole that determines its size and strength?

– Yes, but not entirely. A black hole is surrounded by much more than humans have discovered. There is indeed a connection between the inner and outer parts that we always need to consider when seeking answers. It is precisely in this compressed connection that we find answers about what happens to all the contents when a black hole dies ... according to humans. So if you connect the event horizon with the strong particles that create lightning bolts in the interior, you will also find answers about what happens to the contents after a lightning strike (clue).

THE CONTENTA LINE

Stephen speaks:

We have also discovered a gravitational force that can be highlighted both inside and around a black hole, which some people are still unaware of. We call it *the content line*, and its nature is abstract.

The content line is, as mentioned, an abstract concept. It is a line that is not visible because we cannot see it with the naked eye, but it exists within the creative process we discussed earlier. In this creative process, a linear relationship occurs, which we won't elaborate on here, but this linear relationship is distinct because it can separate parts and matter due to its enormous gravitational force.

However, a black hole itself cannot break the content line. To do that, it first needs to transform matter into a stronger material, and it is this creative process that humanity has only recently begun to touch upon. Therefore, it is crucial that the content line exists and can be broken from within with immense strength. This is also how a black hole can change its nature.

It is through the content line, which has undergone a splitting, that the process of creation occurs, and it is what we want our brilliant researchers to continue studying now. If you do this, you will also find the explanation for how a black hole can create and process the material it consumes, and then expel everything again when it diminishes. It only transforms what it needs to grow larger, and you will soon discover what this is.

– What does the content line consist of? I ask Stephen.

– The content line is abstract, so is its content. But its abstract nature is only because you cannot see it yet, but you will. The content line contains a force that you will use in future industries because it can be *connected*. Hence, it will function as an early electrical link.

The power in the content line is so strong that it can also be seen as a distiller, which you will come to understand and further develop when you get there. And because the content line can be extended, that is also what you will do in the future. After that, you will further develop the power so that it can serve other purposes as well.

The power also consists of an electrical guiding line, hence the name "content line." It is the one that directs the power and ensures it always ends up in the right position. But to do that, it needs to consist of a determining force, and that is where neutrinos come in. Not that the power consists of neutrinos, but you will apply the power in industry according to the actions of the content line, with neutrinos carrying the power, the current.

– So is the power more of a technique in this context than the strength of the content itself?

– Yes, it will be that way for you once you have developed quasar technology (we will discuss quasar technology after this).

– How do we find the content line?

– You find it by searching in the source that releases (clue). It is where they are linked together, and they are linked because they always cooperate. So search for the source that releases and the cooperation that happens between the two – the release and the content line. But to find it, you first need to create a new system that generates a new perception. It is a line that is active but not yet visible to the human eye.

– How can we then know it is there?

– You will see it through a new telescope. It is indeed already in place but only as an accompanying instrument, and it is what we need to integrate with today's technology.

-How can the instrument help us see and understand the content line? And what instrument do you mean?

– You cannot do that now, as the instrument only exists on paper today, but researchers have finally begun to understand that we also need to look beyond the external. So when that blueprint is put into use, you will be able to see and understand why a black hole attracts so much as it does. It is the one that connects a black hole's gravitational force, and everything will be illuminated in the future.

The instrument, as mentioned, is an accompanying part and will expand your vision exponentially. It is also designed to change a picture's ability to *perceive* (clue), which allows you to see beyond what is real, if we can call it that. Our researchers already know what this means; now we are just waiting for a result.

– Why do you call it an instrument if it is an accompanying part?

– We call it an instrument simply because it will act independently in the future. So, it is its own entity, but it can still be connected as an accompanying part.

Archangel Metatron speaks:

We cannot reveal more about this immense force because it is forbidden for us to do so. But we hope it has given you a start to continue your work. Because it is not what it accumulates that is important, but *how* it does it. And no, the content line and Hawking radiation are not the same thing as they represent *difference points*, as we call them. There is indeed a difference between Hawking radiation and the content line.

For example, radiation is a process that emits electromagnetic energy. The content line consists of the gravitational force itself, the energy present there without blocking or disrupting magnetism. Because that's what radiation can sometimes do, it can block magnetism without touching it, and this is where our line comes in. It can actually enter and break a beam once it has transformed its material into a stronger material, which radiation cannot do according to us.

We understand if some are hesitant now, but we are only talking about things we know exist and operate in the universe.

Think of it this way: Not all scientists have all the answers, but we have many of them, not all but more than you. Therefore, we cannot rely solely on human knowledge; to learn new and correct information, we must reinvent ourselves. There are scientists who need new glasses because they are stuck in the old 3D thinking. Thus, some researchers will oppose and not believe our words, which is perfectly fine. But all new things must start somewhere, and if we never try new paths, we will never gain new information.

So find this power source and trust that it exists, and you will also understand how everything can accumulate. You will use the content for our electrical components in the future, especially where we need to highlight forces that attract and fuse matter. We mean those components that have reached a high level of activity so that they can transform a material into a stronger material, as in a black hole.

Moreover, it will also help you when you need to highlight other foreign objects in the universe. Because if you understand the dynamics of power, you will also understand other phenomena in the future. So, combine your efforts and try to study the content line – how strong the gravitational force is in the inner and outer atmosphere. Then perform a transformed procedure, what happens in a black hole.

However, when we talk about quasars, it is their control over the strength that interests us. Because, as mentioned, through this, you will develop your own quasar technology.

QUASARS

Archangel Metatron speaks:

A quasar is a primordial star, a supermassive black hole located at the center of a galaxy. It is surrounded by an accretion disk where matter radiates around and spins at extremely high speeds. If we were to translate its capacity into a human factor, we would call it *hubris*. The quasar accumulates everything in its path and has an insatiable hunger and drive. That is why it can become enormous. Its ability to grow so large is due to its magnificent gravitational force, which allows it to choose what it wants to consume to grow larger and more powerful.

This gathering of various energy-rich contents, along with its enormous gravitational force, is what humanity must now research further and use this combination of different nutrient sources to achieve more energy than you already do. And since a quasar's life arises from energy-rich substances, which provide it with an energy-rich mass, you can adopt the same technique in your daily life, but only once you achieve the same technical prerequisites.

– Why do we need a primordial star in the center of our galaxy? I ask Metatron.

– The existence of this primordial star was necessary; otherwise, we would not be able to create a galaxy. The origin was based on cohesion. Without this original cohesion, our galaxy could not have developed.

– So, the quasar was needed for a galaxy to form, otherwise, it would fall apart, correct?

– Yes, but not completely. A galaxy can also be formed from other elements, including quasars. But then something entirely different forms, although it is still a galaxy.

– What does it form then?

– A spiral, long like a serpent, but in this case, it is the head that controls it. Therefore, humans can also see other galaxies that do not look like a traditional galaxy but are still galaxies. And as long as there is no quasar in the center, the galaxy cannot hold everything together in the same way that our galaxy does. Thus, some take the shape of a spiral. That's why diversity in the universe is so important; otherwise, we would not develop.

A quasar can also control its galaxy to prevent the formation of too many stars. It can, if necessary, slow down the burning process so that the galaxy can live a little longer.

The quasar is also known as a crusher, a crushed mass. It has managed to accumulate so much nutrient-rich content during its journey that this is what has made it so strong and ultimately created an enormously powerful energy channel.

The quasar is so powerful that it can destroy almost everything in its path, but destruction does not always have a negative impact, as it can also lead to other things as a result of this destruction, which is also the quasar's main purpose – to travel, destroy, and then disappear so that a resurrection can occur.

"From the old grows the new," as Archangel Metatron often says.

Therefore, when we direct our attention to a quasar, we may feel that it resembles a malignant tumor darkening the brilliance of the universe. But the universe does not follow rules that say what shines is good and what is dark is evil. Everything is equally important here, and the strength of the universe actually comes from the color black. It does not matter if something is white or black. Everything interacts; otherwise, the universe would extinguish.

So, as humanity deepens its knowledge of the quasar and its composition, you can provide yourselves with much better energy fuel in the future. But be warned, people! It is a dangerous path to tread as long as we still have some egocentric souls on Earth, as it could also lead to weapons being used against each other. But in the right hands and under the right conditions, the quasar's drive is a good investment for the future. And since the technology is still some time ahead, and if it were to fall into the wrong hands, we cannot reveal more than we do.

The quasar's life purpose is to destroy to create anew.
It breaks down to then add and increase capacity.

QUASAR TECHNOLOGY

Archangel Metatron speaks:

A quasar collects matter from the outer regions. Different parts merge together through the gravitational force present, allowing it to continue rotating. Consequently, a quasar contains decomposed matter that has been fused with other components, and this is what makes it so powerful. It illuminates everything in its path, and all the energy it generates allows it to even alter the universe.

A quasar has a controlling force that some scientists are already familiar with. But did you know that it can also release and gather more matter depending on whether it feels heavy or weak – full or satiated? It does this through explosions. Through these explosions, quasars can release matter when needed and, in doing so, contribute new material to other areas where there may be too much or too little of something, shifting the volume to different parts of the universe.

That is why we refer to them as our gatherers and expellers. You could say they are like security guards with accompanying security measures, some of which might seem intimidating because the power can be perceived as such. But for us, it is a natural part of the universe.

This is the part we want to talk about now and find so exciting because we know that if you choose to pursue this path, you will use this quasar technology in the future. It is a technology that will enhance both human perception of the universe and understanding of its power.

There are three steps you need to grasp when developing quasar technology:

1. It collects matter and becomes full.
2. It expels matter.
3. It collects everything again. This way, it restores everything and can start anew.

A quasar can stop on its own when it feels full – satisfied. It has achieved its goal. The matter that the quasar then expels as a laser beam can actually be reused by the quasar. It is when it starts to collect everything again that it can restore and reignite its inner power. This restoration and reignition is something humanity can use in the future as a way to drive movement forward – restarting with new energy when the old energy has ceased. We could call it an on-and-off switch. It is not just the technology itself that we will use, but also the tempo and, most importantly, the transition that happens when we reignite everything that we need to learn more about.

But to keep everything running, you also need to improve your knowledge of fusion combustion. Therefore, two factors will be especially important for your future research:

1. Ignite fusion combustion more effectively than you currently do.

2. After that, the quasar will expand your knowledge and outward capabilities. The quasar's extent and contents are much greater than what humans can currently measure. Therefore, your calculations will also change in character here.

You can also explore future energy discoveries in the same way you do with neutron stars, but with much greater power. For just like neutron stars, quasars also accumulate surrounding matter. So, when the pressure increases, it can convert gravity into magnetic energy, leading to kinetic energy when the beams are shot out. When you can see the similarities between a quasar's and a neutron star's internal and external actions, even though their extents and strengths differ significantly, then you will have made significant progress!

This is a technology that you will also use in industry to extend certain processes. But most importantly, you will use it in the future automotive industry, and that's when you will spread your knowledge of quasar technology. As you can see, there are many great opportunities in the future with quasar technology.

However, we also want to issue a warning. When this technology emerges, some may try to master it on their own in their country, which should, of course, not happen. Everything we do should always be shared equally among everyone. We will prevent this, as no country should have priority. Therefore, we want to briefly explain how you can further develop quasar technology.

How we develop quasar technology further

Everything will happen in a machine that you create yourself. It will be a steam-powered machine that will provide you with the results you aim to achieve.

– How can we develop quasar technology using steam? I ask Metatron.

– For steam to be applied along with the shield you will create with quasar technology, it is essential to work with something that can penetrate. What we mean is that to develop quasar technology, we first need to use materials that are easy to work with and easy to develop further, and steam is precisely what we can use for this. We can easily add other elements to the steam to enhance the effect.

– So, will quasar technology use steam in the future?

– No, not exactly. Steam will not act as a power source since it already exists. Steam will only act as an enhancer, as it needs to pass through various stages for the power in quasar technology to increase. Initially, the power will not be very strong, so you will need to develop it.

– How can steam increase the power? I don't understand.

– It will assist the quasar's contents to function as an enhancer. When we enhance one part, it positively affects other parts as well. So, it will be steam based on positive elements. It is the positive elements that increase the power, not the steam itself.

But to add new elements to the steam, you need to first understand what you are looking for when expanding the power in quasar technology. This is where you will face a challenge in the future. Therefore, our advice is to examine what integrates with a quasar at the speed it operates. What elements cause a quasar to increase in strength? However, we don't want you to view each element separately. See them as a welded chain because that's where the power lies – in the chain. It becomes an extension of power, and that's what we previously mentioned.

Look at steam locomotives. They might give you a clue about the steam itself and how it moves.

The quasar technology now rests in your hands to develop, use, and apply this phenomenal strength, this phenomenal technology, in your daily lives. So, listen to those studying quasars and provide them with the resources they need so they can continue observing and mapping and creating a new *ring* for everyone to see (hint).

We have also already placed active observers in various programs to review "their" activities and contact us if needed. This is to ensure that those who do not share their knowledge with others are noticed and to prevent them from creating new future plans.

CHAPTER 9

POWER PLANTS

NEW POWER PLANT

Archangel Metatron speaks:

In addition to the magnetism we will develop in the future, which we'll discuss in the next section, we will also create an entirely new power plant. To explain this, we need to revisit the neutron star.

For a neutron star to gain momentum, it must both harness and generate power, and it is this synergy that is necessary for you to start up a new power source. This is a technology you will use in future power plants due to the speed and variability of neutrinos.

But a neutron star is not just about immense internal power; it also acts as a balance carrier. To do this, it must first achieve these properties. Much of the power generated is not created in isolation; it collaborates with other parts of the universe. Therefore, we also need to understand what surrounds a neutron star, such as the magnetar.

It is also this outer material surrounding a neutron star that some researchers have already discovered parts of, and thus they have nearly found the power circulating around a neutron star that you will use later. At the moment when the outer and inner matter meet, things begin to happen. What happens then is that the power increases. So without this external and internal collaboration, a neutron star cannot achieve its powerful effects or emit bursts, and it is this phenomenon and its surrounding aspects that are now capturing researchers' interest. So keep at it, as you are on the right track.

This collaborative method will be used in the future. It is a powerful synergy that needs to happen now for you to get another power plant, consisting solely of magnetism, up and running. You will create these methods based on neutron star technology. You will use magnetism in the same way, thereby replacing current power plants.

It is said that some power plants should not remain on Earth because they do not contribute to Earth's well-being. However, our advice is to keep them as you will revive some of them in the future, but as better and cleaner power plants. So, stay where you are but don't get rid of things! Let time take its course and continue developing where you are now, and you will soon understand what we mean.

These new power plants will make big headlines in your newspapers when they come into play. The magnetic capabilities of neutron stars will be used worldwide in the future to manage your more nutrient-rich resources. Therefore, we recommend that you now turn your attention to removing parts of your power station that are not beneficial for the future, while keeping what you can use.

This is also what we previously discussed regarding neutron stars and neutrinos. That's why we're not elaborating further. We mention it here only because of the new power that neutron stars will release in the future, and that's when you'll understand. So start with the basics, and when this new power arrives, you can begin to develop your power plants based solely on the internal concepts of neutron stars.

This is a technology you will further develop after the car has made its breakthrough.

NUCLEAR FUSION

Nikodemus speaks:

Another aspect we recommend, which will positively impact your energy flow, is fusion. It is a technology derived from a star, and it's the same technology we use when we increase our power.

So what is nuclear fusion? When a star forms, it primarily starts with a thin cloud of hydrogen being pulled together by the immense gravity of the universe. Once the star reaches a certain size, it begins to undergo nuclear fusion, and this is where it gets interesting – when the star starts converting its various elements. The intense heat and pressure inside the star force atoms to fuse together, forming larger and heavier nuclei.

This process produces different elements at various stages, including magnesium. It is through the carbon dioxide combustion occurring in the star that we also get magnesium.

Fusion is the process that drives all stars, including our own Sun.

We could continue discussing how atomic nuclei fuse and release energy, which is the basis of nuclear fusion. But that's not our main focus here. Instead, we've chosen to concentrate only on one element formed in a star during fusion – magnesium. Magnesium will play a leading role in fusion combustion in the future due to its igniting properties. Therefore, humans need to improve their understanding of this stellar technology – how fusion occurs.

Think of it this way: Carbon dioxide combustion is needed to produce magnesium and oxygen. What happens next, once we combine them, is a transformation, and this transformation must be implemented before we can use future magnesium – not the magnesium of today. Once you've learned this and integrated a new type of nuclear fusion into your future technology, you can apply similar principles to your machines. When you then combine this with how a neutron star functions, you'll soon understand what we mean. This is where the movement comes in, among other things, through the fusion used by neutron stars.

Thus, magnesium will always be a key factor in our future production processes, as without magnesium, we cannot start or ignite an energy flow.

- The star transforms matter into what we need in the generator, such as magnesium

- The neutron star contains magnetism and creates movement. The movement captures surrounding matter and rotates it.

Fusion should be used in the same way as in a star, to transform and create new matter, not to combine existing substances to increase capacity. The key here is not the strength, at least not right now. It's the creation of new matter, the transformer that is important.

By studying stellar technology during fusion combustion, you can create future fusion methods. When you then use a transformer, like one that converts carbon into magnesium, you can utilize the entire energy flow to, for example, get a car moving. This will be your first task – how to achieve a better-performing fusion fuel.

– Don't extremely high temperatures need to create nuclear fusion, and wouldn't that negatively affect the magnesium? I ask my father.

– As you advance in your development, you will also discover a new way to manage your magnesium. What we advocate should be used solely as an energy enhancer, almost like an ignition. Once the energy level has increased, other substances take over, those capable of handling high temperatures, just like in a star, and this is where magma comes into play (we'll discuss magma energy after this).

– How can we incorporate magma into nuclear fusion?

– We should not add magma or magma masses. It is the associated steam that we advocate, but not the part that has a negative impact. There are many substances related to magma and volcanic eruptions that are not yet perceptible to humans but can, if handled correctly, initiate a fusion.

So, it will be a steam based on sodium chloride, which will also find a better place in the future. Sodium will form the foundation, but everything will then grow on its own (hint). However, you can also use magma as a heat-resistant effect.

Therefore, magma and magnesium will have a significant impact on our research groups in the future. It is also their job to address this so that we can progress on this major issue. But we know there are researchers who complicate things more than necessary. Sometimes simplicity is enough to solve big problems. We don't always need complicated, highly sensitive formulas to solve tough issues. It's about finding the right formula. So don't give up. We foresee that nuclear fusion will bring you good news in the future in this energy-enriched field.

Fusion is indeed the future's tune, but like much else, there's a lot to address first. Take the time you need, as it is crucial that the process and its elements are managed properly; otherwise, it could be more harmful than beneficial. It is, like in a star, a highly dangerous process to work with.

We are aware that there is already a nuclear fusion reactor on Earth, which is naturally good. But that is not the one we want to highlight in our book. We process our part in our own way, based on the fundamental knowledge we have in the spiritual realm. Therefore, we cannot rely on human talent alone; we always trust our own.

MAGMA ENERGY

Archangel Metatron speaks:

How can we then use magma in our future energy system? But above all, how can we extract magma from Mother Earth when it is so hot?

To harness magma energy, you need to dig for it. But to do this, you need a drill, and not just any drill. It must handle both the amount of earth it has to penetrate and the heat present in the magma. However, since you have not yet reached this technology, we recommend that you instead allow the Earth's crust to lift this energy flow. What humanity does not yet know is that it is also possible to extract energy from magma without having to bring it up, which can be done by starting a small volcanic eruption yourself.

How do we do this? We do this by understanding how emissions work in a volcanic eruption. It is by releasing a little gas at a time that humans can create their own small volcanic eruption. Just enough to lift the portion of the magma you need. In this way, you can extract magma energy from Mother Earth without causing any harm.

But in order to use magma correctly in the future, we first need to understand what it contains, but also what it does not contain. To do this, we need to release certain parts, certain particles, that consist of heat energy, and then leave the remaining content as it is. This way, we do not deplete all of Mother Earth's magma resources, but only take what is important at that moment. The rest you will learn about later.

If you use this technique before you have learned to manage and take care of Mother Earth, it will not take root. Your chances of starting something good will slip through your fingers. Therefore, you must first learn to handle what you extract without adding chemicals before you can move forward.

– How do we release particles? I ask Metatron.

– We extract the particles that generate heat in the magma and then place them in a heat container. The heat container will function similarly to today's batteries, in that it can store and release energy. Trust us, it will work as long as the particles are contained in a heat container.

The process of extracting particles from magma is already partially underway. It is more about how we can manage and further develop the same technology for use in more places, for example, as a fuel in the future. And for future heat containers to work, we need to add more magma than we currently do; otherwise, it will not work from an energy perspective. Therefore, initially, and before humanity fully understands its concept and function, they will remain relatively large. But the size will decrease as you become better at extracting what needs to be extracted.

It will be a container, a generator, that can take and release heat particles.

Let's take an example and return to Mercury:

Think of an orange, where the peel represents Mercury's surface and the rest of the orange is its inner core. Then ask yourself: How can we use Mercury's way of handling its internal and external heat to create a heat container? And can we find a material that can handle the magma we previously mentioned? Additionally, if we have a warm mass, for example in a generator, do we need, just like Mercury, a strong magnetic field to protect the contents? We answer yes to all these questions.

But in order to develop a magnetic inner core, it needs to be large, at least initially, just like Mercury. All development takes time, but it is also what will then take you to other tasks. Therefore, it is crucial to study Mercury's inner core and its associated magnetic field, as the heat generator will need to be close to the heat you will produce in the future. Additionally, it will create its own strong magnetic field, but in your case, a strong magnetic attraction.

Furthermore, we can ask another important question: Why does Mercury not get pulled into the Sun's strong gravitational force? In other words, how can we create an outer shell strong enough to withstand the universe's powerful gravity, just like Mercury does? Moreover, how can we avoid the gravitational force we subject the generator to either being destroyed or completely disappearing? Something for our future researchers to tackle. This is how we want to emphasize Mercury as a help in increasing your knowledge.

By adopting Mercury's method, using Mercury's system, we can thus provide better internal protection in future heat containers, generator blocks, so that the heat does not leak out. Therefore, we will establish a completely new technology in the future; otherwise, no improvement will happen (we will discuss this further when we talk about magnetism).

– What do you mean by "using Mercury's system"?

– What we mean is that Mercury's system does not tilt, its axis does not tilt. Which leads to, despite the planet slowly moving around and near the Sun, there is ice on Mercury. But how can a planet so close to the Sun have ice? How is that even possible? And how can Mercury's front side be hot, while the back side can become hundreds of degrees below zero?

Because Mercury has no tilt, its north and south poles are always in the same place. This means there are no environmental changes. The cold remains cold, which allows ice to be preserved in Mercury's thin mantle. By using the same technique, with a constant negative pole and a constant positive pole, we can achieve the same movement and result. This means there are no changes in temperature.

We are not referring to today's positive and negative, but to a stronger, stationary position that cools at the north pole or warms at the south pole. In other words, instead of positive and negative, we create positive poles and negative poles, and so on.

We know that positive and negative poles have already started their journey on Earth. Therefore, we thought it was a good idea to use Mercury as an example, as we want you to further develop your positive and negative poles to function the same way Mercury does – to be stationary but still act as protection for the content they have, to protect and preserve its temperature. Additionally, we need something that can link heat and function as electricity, which can partly come from the Sun but can also come from magma power.

– Will magma replace batteries then?

– It will not work exactly that way, as magma generates a hundred times more energy, if not more, than a battery can produce or handle. So it cannot completely replace a battery since magma cannot conduct electricity. But this is not something we will cover here.

In conclusion, we want to add:

We should also mention that magma is already used in many ways today, where it is broken down into various functions. Among other things, stone blocks that are then used in stoves and so on. This is a technique where the magma is first allowed to cool down and then broken down into pumice stone, and this is exactly what we will do with the heat in the future.

The difference lies in how we learn to ignite the pumice stone and how we can use it to generate heat in our container. Therefore, we will see more of this in the future, even at some research stations. But to ensure that, for example, Iceland is not overrun by people as more and more start using magma, it is fundamental that you first learn how to start a small volcanic eruption.

– How is it that, despite many already analyzing magma and volcanic eruptions, some still do not understand or know its functions?

– We never allow humanity to gain more knowledge than you are ready to face. It would simply create a new rush about who got there first or who knows the most. Let's say it would save Mother Earth, if we were to give you a secret code for a better fuel right now.

– Wouldn't that benefit humanity and Mother Earth?

– Absolutely, but it would only lead to a short-term change, and as you know, we always work from a long-term perspective.

– Finding a better climate-controlled fuel is a long-term perspective, isn't it?

– Absolutely, but it is also important that humanity keeps up and develops at the same pace.

We mean: Imagine if a solution to solve all the world's problems fell into your lap, we want humanity to also work from a long-term perspective, which is not always done according to us. There is sometimes a competition about who was first on the moon or which country is best in various areas. In our view, this is just egoism and prevents humanity from working from a global and long-term perspective. So before we give you more knowledge that you can further develop, for example, volcanic eruptions, we need to first change humanity so that you work from a global perspective.

– Will you then tell us how to reach a better solution?

– We will always be here to support you, but to give humanity a ready-made recipe for solving all worldly problems right now would be wrong of us. Humanity must solve its own situation, which you will do. We only talk about magma energy to guide you by saying that you are on the right track with that issue. Consider it as a kind of confirmation. Therefore, we want you, who study future fuels, to also study magma's powerful energy levels.

But listen and hear our plea! This medium, this technique, should only be used either as fuel or for heat, nothing else! If you do otherwise and defy our words, we have been given permission by our highest Lord to intervene and change your starting point.

Summary:

We convert energy – heat is generated, which translates to energy particles (components). The particles are collected and used as heat energy in a heat container powered by coal. Coal, together with magma energy, drives forward movement, such as in a car. Coal will be replaced once we have improved our fusion fuel (nuclear fusion).

Magma energy heats similarly to the Sun but has a more sustainable effect when it comes to transportation – just as magma moves underground. It is simply a more sustainable form of heat.

The reason we do not see using magma as a threat to the future is based on two factors:

1. By understanding magma as an integral part of a volcano, we can actually prevent some eruptions in the future. This means that by using magma, we reduce the pressure

inside the magma chamber, which in turn reduces the need for the volcano to erupt.

2. Magma can never run out in the same way that coal and oil can. This is partly because it takes a long time to reach it, and it is somewhat more complex in its context. Additionally, the energy it emits is so powerful that humanity only needs a fraction compared to coal and oil.

Therefore, we do not see any danger in humanity delving deeply into this subject. Mother Earth has also given her approval for this, and there is plenty of magma that humanity can use, she says. However, the technology will not develop until humanity has learned to take better care of Mother Earth.

MAGNESIUM – FOLATE

Nikodemus speaks:

Magnesium is one of our most important elements. It has been with us since the universe created its first mass, when both life and energy were formed. Therefore, magnesium is spread almost everywhere, even in the Earth's core.

Magnesium also creates a free flow in the universe and provides all the parts needed to assemble something, such as a mass in the universe. Therefore, it's crucial to understand magnesium's role, both as a cohesive and integral component. Magnesium will always be a building block of the future because it functions as a guiding nutritional chain in the universe, the spirit world, and here on Earth. What do you think would happen without magnesium? How could we exist on Earth then?

Without magnesium:
There would be no life or energy on Earth.
We could not develop.
We could not move in our physical bodies.
We would get sick.

Other fundamental elements necessary for life include folate. Both magnesium and folate are substances that early on could integrate with other sources in the universe, acting with full strength like a star. Therefore, magnesium, like folate, was a key source when our first star was created. This is why we highlight magnesium as the cornerstone it truly is – a real survivor and a strong building kit. A nutritional source on which our entire foundation is built.

– Isn't carbon our strongest element? I ask my father.

– True, carbon has also been present in our atmosphere since the beginning of time. But we don't see carbon as our strongest foundational element. Just because an element came before another doesn't make it the strongest foundation. A strong foundation is not always about its origin. For us, a strong foundation is what remains when everything else falls, which is more important than the creation itself. A strong foundation is formed through various interactions, and without these interactions, we couldn't build a strong foundation.

Let's take an example:

Suppose we build a deck with boards and nails. Together, they form a strong structure, but we believe it's always what survives that forms the foundation. What do you think remains after the deck has decayed? The nails. Thus, the nails (magnesium) form a stronger source, and the boards (carbon) a strong foundational base. Carbon, in many respects, decomposes more easily.

– Doesn't magnesium also decompose easily?

– Yes, today's magnesium – but not the magnesium of the future. We won't delve deeper into this now. It is what it is, and not everyone needs to believe our words. The deck and nails were just an example to show that the strongest survives and forms the foundation – the one that remains.

Magnesium and folate thus became important nutrients for all human life, providing the strength needed for both our soul and physical body to live and thrive on Earth. Without folate and magnesium, we wouldn't have strong building blocks. This is why magnesium is our greatest scientific source and will always be. It is also what our researchers will continue to study. It is a very strong candidate for the future.

OTHER FUNDAMENTAL ELEMENTS

Archangel Metatron speaks:

There are few additional fundamental elements that we briefly want to highlight now. These elements will have a direct impact and will be crucial as you learn and further develop this concept. New chemical notations will change their structure in the future, which will also advance your chemical map. We begin with helium.

HELIUM

You will create new helium from the old, which will open up new possibilities and uses beyond just balloons. This means taking the existing composition and extending it.

However, start by introducing your new helium as an artificial variant – for safety reasons. This will then open up new opportunities for your chemical introduction model. When you later mix it with other components, it will take you where you need to go. So learn to harness the new power of helium and also use it in the industrial world. Helium will be useful there as well. Therefore, you also need to produce a new type of nitrogen that can be used with your new helium supplement.

Hint: Roll forward in current days. What makes it roll?

Cleaner Water

Helium can also be used to purify water and provide you with another water source. Therefore, it is crucial that you do this; otherwise, more people will die, either due to water scarcity or polluted water, especially in the countries that have been hardest hit. Start there!

To transform our drinking water and restore it to its original purity (holiness), we also need to make new calculations.

Ulrika: I am now writing down what comes to me without having knowledge of either the order or the content.

U^2 (not uranium) – Hydrogen, H^2 – Mg – Helium

Hint: We can bring much to life when we learn to use helium in a better way.

It will become a future concept where we highlight the art of helium by adding other components, which will start after the corona has subsided and people have returned to their daily lives. Then new students will gain knowledge of *how helium can actually transition and change so quickly* (hint).

Once they understand this, things will move quickly. The reason you have not yet been given this information is due to the amount of information that "they" have sealed off.

SULFUR – OXYGEN

Sulfur and oxygen will be used together in the future, for example, in cars. Both will be part of the new gas you will produce shortly. But it will be an artificial sulfur that will be used in the near future. It is a new sulfur that does not spread evil. It is a sulfur that will be used together with oxygen because, after some research, they will cooperate, but in a completely new way.

Some say that sulfur and oxygen can never be combined, but in this context with your new artificial sulfur program, they will cooperate. When this is revealed, you will be immensely delighted with this discovery.

However, it is a dangerous concept but one that will be solved by our scientists, those studying the chemical structure's content within its chemical notation. Therefore, your students will raise the issue of this shift – about sulfur and oxygen. So listen and heed their words, and help them develop. It is important that you do this and that you trust their words. They are here to help. And do not laugh at your students; they are all fantastic but need guiding help and support so that a continuation can occur.

PHOSPHORUS

Phosphorus will also become one of nature's enhanced elements in the future. Therefore, it is our desire that you continue researching how it can be activated and move forward. Phosphorus will also be used in industrial development, where it will be used as a gas product. However, it will be a gas that only touches, not destroys. So continue to develop the gas, including phosphorus.

It will illuminate your path as a guide, not as a part of it (clue).

Phosphorus will also be present in your future kitchen, where you will further develop the technology that is already ongoing there. Therefore, you need to develop your instruments, as today's do not reach all the way, which you will soon discover. Once you have done that, you can also handle your new phosphorus in a gentler way, and if you follow the rules in place then, no one will be harmed. So be careful but dare to develop, is our advice. For phosphorus will be discovered and used in an entirely new context, and when you do, development will take off like a rocket.

NEW ELEMENT

God Speaks:

A completely new element will also emerge, which is also created during a star explosion. But it is not from a red supernova but from a star that many are not yet familiar with, but which you will witness in the near future.

Clue: Some astronauts will understand what we mean because they have already traveled to the star you are now to return to. You will use and further develop this fundamental element, as it will positively impact your future research.

However, it is not a finished element; it is a technology that you must create yourselves. It is a technology, a production, which will initially require its person. But do not give up, for you have plenty of intellectual souls on earth. And you will succeed, and when you do, you will understand why and where you can use this fundamental concept. It is a fundamental element already used by many other galaxies in the universe.

It is a technique to use and formulas
to transform
to create and produce this new element without
taking from Mother Earth.

Consider it a kind of replacement for lithium, as you should no longer use lithium in your batteries. Lithium, like many other things, needs to remain in the beautiful womb of Mother Earth. Therefore, the use of cobalt and lithium batteries will cease entirely, as it is a much more dangerous element than you are currently aware of. Very dangerous products can be produced in the future if you do not stop using this concept. A warning is therefore issued by me now, so listen to it! We do not want more motor parts to explode due to ignorance.

God signs this future notice.

Archangel Michael Concludes:

Now that we have gone through the parts that we need to improve and understand, we will finally enter the magnetic world. For it is there that our future lies, as it is what sets everything in motion. It is also where we create levitation.

PART 2

KATHREMES – MAGNETISM

CHAPTER 10

MAGNETISM – ELECTROMAGNETISM

MAGNETISM – ELECTROMAGNETISM

The angels speak:

What is magnetism and electromagnetism?

Magnetism arises from electrical forces, where, for example, two magnets attract or repel each other. Magnetism means that electrically charged particles are in motion, but to keep everything together, we need a magnetic field.

It's like when two people hold each other's hands. The movement, the convergence, that occurs then is created by an automatic magnetic motion. And this is exactly what we are looking for in our newly manufactured cars – an automatic magnetic motion that can replace the electricity consumption we have today.

Electromagnetism depends on magnetic and electrical forces, which means that there is an interaction between electric fields and a magnetic field. Electromagnetism is thus the part that unites electrical and magnetic phenomena and constitutes the inner part of the magnetic world. And it is precisely the inner world of electromagnetism that will be important for us in the future, and that needs to be improved because parts of the concept are incorrect.

HARNESSING MAGNETIC POWER

Kazandra speaks:

How do you identify these inaccuracies? You do so when you reach the magnetism of the future. And to explain that, we've divided it into three parts:

1. To reach the magnetism of the future, you first need to understand why today's magnetism doesn't work (or hold) entirely as it should. After that, we'll go through some methods you can use to increase the power of electromagnetism.

Part 1 is also about working in sections. We always divide magnetism into different sections, where each section has its own candidate, which we call it instead of a student. This gives more respect for the teachings we are there to learn, and it is always how we work. That's why we've come much further than you have.

So, the biggest limitation is that you often discuss the entire magnetic concept, instead of breaking down your work, which would make it easier for you to find the missing link.

We are also more in harmony than you are. There is quite a bit of egoism on your planet regarding who discovers something first. You need to learn to always share your successes and to always work together – not against each other. This was and still is the biggest reason you have limitations.

2. After that, we journey back into the universe. It is through magnetic fields and magnetic bubbles that you can further improve your electromagnetism.

Part 2 is more about the actual course of events, when and where your magnetism decreases. And why does it do so? It decreases because it does not have the power, the inner strength, that it needs to be long-lasting. For magnetism to be long-lasting, it needs to be used. It is when you use magnetism more than you already do that you find more limitations from which you can learn.

So part two is about using your magnetism, because that is what will develop you. For if you do not use your ongoing magnetic projects, they will stall, and you will also limit yourselves.

3. Then we conclude with solar magnetism and the transformer you will build in the future.

Part 3 is about getting your motion components moving. A motion component is a sustained motion. Something that is strong and maintains speed but can also slow down if necessary. But to do that, you first need to create some kind of remote control – something that can turn a magnetic ability on and off. Because this is exactly what you lack. You lack remote control in the way you manage magnetism.

Remote control is a tool that will sort out the situation you are in today, and when you reach it, you will surpass the limitations in today's magnetism. It is a remote control, a technology, that was based on one of our stars and which we call *the magnetic aspect of the star.*

This is what we will go through now and what will follow us all the way to the end of this second part. We begin with point one: why today's magnetism doesn't work as it should.

ANCIENT CONCEPTS

Archangel Metatron explains:

To understand why today's magnetism doesn't work as it should, we need to go back to the time when the Romans built their empire. It's actually there that you will find future solutions to the magnetic aspects that aren't functioning properly, because even then, they were on to what we know today.

Hundreds of years ago, in ancient Egypt and even in the Roman Empire, they had already begun a serious attempt to understand the movement capability of magnetism.

Even the constructions of the Roman Empire were built by scientists who early on learned the art of carrying, lifting, and moving forward. But as we know, there was no magnetic capability at that time. However, they examined the basic conditions for magnetism – movement. For back then, movement was something completely different than what we have today with cars and high-speed trains. At that time, they could only illustrate movement by transporting and lifting. So what they discovered was a new way to transport stone blocks – they used a rolling technique.

They also understood early on that a few men were not enough to lift these stone blocks. So what they did to solve the movement in height was to combine both techniques, that is, both a forward movement (rolling technique) and a lift, which was a form of early magnetism. The scientists of that time then concluded: "To understand how we can harness the power of movement (magnetism) and weight (gravity), we need something beyond the ordinary," these men said. So what they did was place people in a line (neutrinos), where each one would push forward these stone blocks.

But to lift the stone blocks even higher, they needed a movement that was higher and stronger (the power of the neutron star), hence they created a height of forward movement power. And by placing people in a line, even when it was uphill, they applied lifting power. Then they developed the same lifting technique, but that's not something we will delve into now.

Now, this was just an example. The Romans had more and more advanced techniques than this, but we believe you understand.

Now, many might think this has nothing to do with magnetism. Yes, we say it does. We mean that without all those people standing in a line, it wouldn't work. And that's where we want to get, even with our future magnetic ability. To increase the ability of movement forward so that you can then lift upward. For how will you be able to lift your cars if you don't first have a movement engine that can move you forward?

And no, your future cars will not be carried or hover among the clouds – at least not at first. It will only be an illusion that your car is hovering. The car will actually be lifted forward by the power of movement, but it will still remain at ground level. Further in the future, you will hover higher up.

So try to look back at the technique the Romans used, and when you understand their carrying concept, you can multiply the same thought pattern in your future concepts. And it's actually not harder than that to make magnetism hover. The reason you haven't yet found the missing link, why your ability to fly with magnetism hasn't taken off, is due to those who don't want you to solve this *ascension*, as we call it. Therefore, it's time for us to reclaim the magnetic catch, the one our Egyptians and Romans investigated at that time, and continue where their work ended. There are indeed quite a few inaccuracies there that we want you to sort out.

Our advice is: don't search in pyramids or sarcophagi. Search in the place where everything converges. That is where these inaccuracies have been locked away. Trust us, for we know they are there. So search, and don't give up because it will help you understand what went wrong at that time. And if you are not granted access, ask us archangels for help. For we are the only ones who can grant you a dispensation in this matter. We will then open up a new path for you.

"Reclaim and retrieve the magnetism that has long been hidden," says God.

DISTORTED MAGNETISM

Archangel Metatron speaks:

So if the Romans already explored magnetic ability back then, why haven't there been greater advances in magnetism?

– Today's magnetism works, right? I ask Metatron

– Yes, but not entirely. You've learned some things, but there are still many underlying issues that need to be studied, that you need to investigate because the sources you have aren't entirely correct. We mean: Have you never wondered why magnetism hasn't developed more than it has, when the Romans were already so advanced?

Magnetism has already started its journey on Earth, where, among other things, we have a train that uses exactly what we advocate. So why don't we use or further develop it? Isn't that strange? And why has your magnetic development stalled despite the presence of so many skilled researchers on your planet? Why do you think that is? It's because it has been distorted.

By distorted, we mean the part that doesn't work entirely. The distorted part consists of an internal concept, and this is where we touch on the electromagnetic ability, and it is this that has stalled.

– How do we know where it goes wrong?

– To understand that, we need to understand different aspects of time because it's only in the time you live in today that it is distorted.

– I don't quite understand. Do we live in a world that doesn't really exist?

– It exists, but since we live in multidimensional worlds simultaneously, we also reach different levels of knowledge. And the information you access on Earth has therefore been distorted; it is partly correct but not entirely.

– How come that we live this way?

– Earth exists for the soul to develop, and to do that, we need to be *multidimensional*, as we call it. Otherwise, we wouldn't develop, and time would stand completely still.

– Do you mean that everything we learn comes from the old Earth? That's the feeling I get when we talk.

– Yes, exactly. The old Earth was created first, and it is also the one many still live in. While some souls, like you, Ulrika, have begun a new journey to the new world, and you are slowly transferring to a new time – a new Earth. Hence, your resources are also changing. You simply access new information.

– What happens to the old Earth?

– It disappears, but that's not something we'll delve into here. Therefore, we will now slowly and surely transition to the new Earth so that everyone who has had the sensation also has time to do so.

– Doesn't everyone have time to do that?

– No, not everyone will develop that much in time.

– What happens to them then?

– They will have to repeat the same journey, but on another 3D planet. This needs to happen; otherwise, we cannot follow the ascension that is taking place now.

We return to magnetism:

So it is in the time aspect you live in now that magnetism has been distorted.

– In what way has it been distorted?

We will use baking as an example:

If someone were to change the contents of baking powder, you would no longer get the same good results as you did when the baking powder had the correct composition, so that the cake can rise as it should.

– So the composition has been distorted?

– Yes, you could say that. Because think of it like above. With the wrong composition, the cake cannot rise in the oven, in this case, lift.

– Why has it been distorted?

– So that you cannot develop levitation. But it's only the developing part that has been distorted; the foundation has been laid, as I said. It's the rest of the magnetic ability that you need to further develop. Therefore, to find where it is distorted, you need to first reach the correct magnetism because we are not talking about today's magnetism, but the one that can only be transformed once.

There is indeed one more catch before magnetism transitions into electromagnetism, and that is that it can only be transformed once. We mean that when magnetism is in a so-called dormant state, it can no longer be transformed. Then you simply have to start over from the beginning.

That's the only part we want to highlight now because it is of great importance that you only process magnetism once. It is a purification of electromagnetism that can be transformed multiple times, which we will discuss further in the future. So don't wait. And if it doesn't work, then it doesn't work, and as I said, you have to start over.

– How do we see that magnetism can no longer be transformed?

– We mean that it lies a bit more at the forefront, so this doesn't concern today's concept. Here, we are talking about future magnetism, and it is this that cannot be transformed more than once. And that it can't be transformed more than once is because it is used up, as we usually say. And if it is used up, we can no longer use it.

– Can't we prevent that from happening?

– You can do that by moving away from your old way of illustrating magnetism and starting anew. Not entirely. The foundation for magnetism, the way it acts, can never be changed. What we mean is to develop a new form of magnetism, and it is this that can only be transformed once.

– What do we transform the new magnetism into?

– Into electromagnetism.

– Ah, okay. Of course. So when we have found a magnetism that can only be transformed once, then we have found the correct magnetism, right?

– Yes, you could say that. And that it can only be transformed once is a barrier that God has placed there so that no one can counteract his internal concept. This then becomes a future module on which you can build the rest of the magnetic concept.

Clue: To achieve the correct amount of magnetism, when we introduce a magnet, we first need to straighten it out. Because it is when we have straightened it out that we can start to control it.

– Do you mean that our magnetic ability is not straight?

– No, not like that. There is no magnetism that is completely straight over a longer distance. If it were, we wouldn't be able to control it. To be able to control it, it needs to act differently, and this is where your distorted version comes in. But to understand where it is distorted, we first need to understand where magnetism is in its incorrect position, and that is what we need to reach and change.

But magnetism is not just here to teach us about the car's movement ability and how we can utilize it. When we have come even further in time, magnetism will also illuminate our homes. We are well aware that this development is already ongoing, which makes us very happy, so keep it up. Because it will, as I said, continue to develop in the future.

You will also get to experience a completely new technology. It will then be the one that generates both heat and technically feasible energy waves. These will be energy waves that, via a magnetic heating system, will guide and heat your entire house. Then the electrical outlets will also disappear. But to get there, we first need to deepen our knowledge in energy waves. Because it is energy waves that will control the future's fuel system, how energy will be distributed and circulated in a machine.

Energy waves are what create electricity, we can call it. That's where electricity is. How do energy waves arise? When magnetism tries to break through a gravity, tries to control a gravity, a so-called collision occurs, then energy waves arise because neither of them yields. It's like they don't want to cooperate but are eventually forced into an unwilling one, which can occur when two positive phenomena meet.

It's in the same way that we can create our own magnetic energy waves, but to create them, we need gravity. So it's not enough with some executions where we add electricity, batteries, or other concepts. To achieve the same mysterious result, we need to use and highlight the same concepts that the universe does.

Therefore, you will discover a new form of gravitational wave in the future. Future researchers will then solve this at their desks. It is also their task to create this powerful energy wave that we can use for the benefit of magnetism.

It would not surprise us if you first create an artificial gravity that gives rise to new knowledge in the future. If you can do that, you can further develop energy waves within magnetism. Because it is precisely energy waves that we will use in the future.

But to understand the entire concept of magnetism, you first need to improve your knowledge of electromagnetism. You do this, among other things, by understanding *how we can purify and divide a magnetism*. But before we get there, we will first briefly explain what an atom is and what the inner world of electromagnetism entails because it is precisely the inner part that you need to improve. That's also where it has been distorted.

CHAPTER 11

THE INNER WORLD OF ELECTROMAGNETISM

ATOM

Kazandra speaks:

- Atom = The smallest component of an element. An atom has a nucleus at its center.

- Proton = The nucleus contains protons.

- Neutron = The nucleus also contains neutrons.

- Electron = Electrons are the particles that orbit around the nucleus, like a shell.

An atom is essentially a fundamental element for other electronic phenomena that enable elements to function and rotate like a star. Therefore, we also need other elements; otherwise, we cannot progress, and what we wish to highlight is electromagnetism. We need atoms, electrons, and more to create a magnetic capability.

The inner world, in the magnetic aspect, also originates from the energy illuminated by our magnetic fields. It is like an inner part of an atom, an inner part of an orange, and so on. So, it's the content that influences the movement of magnetism, and it's the outer layer – the electrons – that acts as a sort of protective barrier.

Inside, there is also something we call *driving forces*. For example, both a magnetic field and our soul consist of photons. And it's the power of the photons that we refer to as driving forces, because they are what makes everything spin. Driving forces are substances that function as motion. These substances can be observed throughout the universe, and without them, the universe wouldn't exist.

So, to some extent, it's accurate when some researchers say that the atom itself could sort out these photons between one another within the circle that exists there. That part is correct. However, in this case, photons are not created inside an atom. Photons are drawn to the magnetic sphere because most of them prefer to surround themselves with an electric sphere, the part they are made of. It's like coming home. The photons that encounter an atom bounce between the atoms and therefore react to the outer layer – the electrons. And it's this reaction that our photons use to create a sacred force.

We call this event *an induction of electromagnetism.*

So, photons are the carriers of electric magnetism because they are attracted to it, and when they arrive, they can use their full potential to initiate an electromagnetic process in conjunction with the atom and its structure. And the further a beam of light, for example, travels, the more the capacity increases, which we can also utilize in our solar cells or other modern technologies.

This is a technique that needs improvement in the future when discussing future cars. It involves allowing photons to initiate the process, similar to what they do in a star. But there should also be a form of ignition present, involving incoming magnetism and other associated elements. Photons, therefore, function as a current-carrying link – they are what keep everything alive. But in this case, it is not the light itself that we are after, the part that is activated when molecular bonds take place. Here, we are referring to the motion that occurs between an atom and a photon.

Summary:

Electrons, protons, and neutrons create an atom. Particles, including photons, are set into motion, which eventually creates a magnetic field. We have chosen to discuss only this part of magnetism, as the rest of the magnetic world is so vast that if we were to write about its size and all its applications, our book would never end.

Photons are carriers of electromagnetic force.

This is the inner aspect of electromagnetism, and it is what you need to improve before reaching the magnetism of the future – the electromagnetism of the future.

THE ENTRY OF THE NEW ATOM

God’s spiritual guides speak:

In the future, you will be introduced to a new atom. This atom will be connected with the new iron that will also change the way you build (we will discuss the iron after this). This is an emergence that has already partially taken place on Earth but has not yet reached everyone.

What do we mean by this? Sometimes, to create something new, we need to renew the entire foundation. That's why we are now introducing a new atom to your world. It's a new, fundamental atom that will illuminate a completely new structure.

This new atom-like phenomenon will be introduced to those working in the construction industry, says Archangel Michael. Archangel Michael oversees everything related to injustices. Many of you have been treated unfairly in this matter because some people have not shared their knowledge. There are those who have already discovered parts of it but have kept it secret from the rest of the world.

So, what does the introduction of this new atom mean? And who creates new atoms? As we mentioned earlier, an atom is an internal process where daily changes occur, both within ourselves and in Mother Earth. An atom always forms the foundation for everything else that will then change. Therefore, we have chosen to place this new, fundamental technology within an atom.

Atoms are the foundation for when we discovered and altered the inner circle of the atom, allowing us to study the changes that actually occur in an atom. Everything always starts from a foundation, both in the spirit world and the entire universe. This means that everything is in constant change. If the universe didn't change, everything would remain completely still. That's why it's important for humanity to first understand where everything originates, as some do not fully understand the synchronization of the universe. And to explain this, we first needed to return to our foundational position – the atom.

Synchronization occurs when new atoms are created in the universe. They first need to establish their foundation – what they are and why they were created – before they can synchronize with other parts. If this process hadn't occurred first, we wouldn't be able to share or influence an atomic change. Therefore, the universe must always integrate a new atom with a new foundational aspect when it is created. This foundation can then be altered so that it can be used in other parts in the future. For this reason, we need to synchronize an atom in different parts for different purposes before it is put into use.

So, you see, there are no random discoveries or results in human research. Everything is created to happen.

– Who creates everything new? I ask God's spiritual guides.

– First and foremost, it is God's source of power that oversees everything. He supervises the content and the processes occurring in the universe. Once the foundation is laid, he releases it to our spiritual firmament. We then have the opportunity to study everything new as well. But as the universe itself grew and learned how to process and create situations, it changed and developed everything at its own pace, in its own way. But there is a clear limitation here: the universe can never change itself until after the foundation is laid.

– How can we continue to change from a synchronization perspective? I mean, is there a limit to how much an atom can synchronize with?

– There is an inherent limitation, and this limitation is important; otherwise, it could be used for purposes that are not of good character. So, everything that can be synchronized with an atom is already built in. When it reaches its limit and can no longer develop, a new atom with a new power will come in instead.

Therefore, it is crucial that we do not create new atoms until humanity is ready to face this new development, which will happen now. That's why we have included another synchronization in this new atom, so it can then act out and be used for more purposes than it was initially intended. Do you understand, Ulrika?

– Thank you, beloved God's spiritual guide, I understand. But isn't the new atom already in place?

– Partially. A new discovery has been made where scientists, as we mentioned earlier, have already claimed it, which should not happen! Therefore, we have now decided to send out a new signal so that everyone can access the new atom and learn from it.

– Do you mean that they haven't yet reached the full content or properties of the atom?

– That's correct, and we have put a stop to this because they aren't sharing the discovery of the new atom.

There is a lot of selfishness within the scientific sector. That's why we do everything we can to share our knowledge with everyone. But since there are those who consider themselves more entitled to the atom than others, they claim it and only act from their own perspective. Therefore, we will now send out different signals at different rates and with different names so that you can then piece together the parts that come to you.

So, listen for new discoveries, because due to the previously mentioned situation, they will come out in different parts. Once you have all the pieces, put them together and create what it is meant to be. But above all, we want you to always share your discoveries – the universe is for everyone!

We are coming in with our strength now partly because it is difficult to understand, but also because it is hard to see that it is actually an atom since it exists far beyond human perception.

- It doesn't look exactly like our current atom, but the structure, the foundation, is the same.

- It will illuminate a better future construction, both for understanding the universe and the content of Mother Earth.

- Its strength is enormous, making it more durable than today's atom.

- Its outer shell will also highlight its weight in gold, for here you will learn to frame and protect things.

Atom – a process is a beginning of a new life.

IRON

Kazandra speaks:

To ensure everything holds together in the future, you will need to develop a new type of iron, which you will create using the power of the universe. This is a universal power that you will harness before the year 2030.

So, how do we build the iron constructions of the future, and what components should be included? To create a stronger iron, we need atoms that can accomplish this. Therefore, you will adopt the introduction of this new atom even within the construction industry. It's an atom that is much lighter yet much stronger. It belongs to a more advanced group and is hard in nature. This is something they are already aware of and have incorporated into their future construction projects.

So first, you will create a new iron by combining your current iron atom with the new one. Once you've done that, you'll begin the next phase – atomic power.

The old and the new atom
will lead each other in the future,
even within the construction industry.

These are two atoms that are almost the same but with a small twist. In our current iron atom, we can actually reach new levels and embrace the idea along with the introduction of the new atom, which creates a new event, altering the atmosphere and generating a new fundamental force. The force we're talking about is atomic power, and in this context, we're referring to the atmosphere – the energy – that is created and surrounds this event. This is the power that you will develop in the future.

Iron will always be a part of your lives, and once you have developed it, you can use it for much more than you already do. So, you too will enter this iron era of newly discovered fundamental concepts, allowing you to build without needing to take sand from Mother Earth, which, as mentioned, some are already doing.

But what they are doing wrong, in our opinion, is ignoring *where* they build, and this will cause them problems in the future. What they haven't accounted for is that this type of atom changes over time. So in a few years, they will realize that this type of element cannot be used as a foundation.

– What happens then? I ask Kazandra.

– Since this type of atom is not constant – it's light and strong, but it's changeable – we can't use it as a foundational concept. What we can use it for, however, is the framework and contents of a construction, but it should never serve as a foundation. Otherwise, your structures will become unstable in a few years. Its changeability comes from the fact that it's dynamic. Not that it can move freely, but it is free to change itself.

– So why should we use it if it causes instability?

– We should use it because it's phenomenal and easy to shape. Above all, it's cheap to extract and use as an ingredient. But you need to think carefully about what you'll use it for. Because if you use it correctly, in the right place and at the right temperature, it's much more stable and sturdier than anything else in the construction industry. But in the wrong place, it won't hold up.

You will also use it as protection for your motor functions. It will require solid protection around it when you start developing your magnetic abilities because there are times when magnetism can erupt. By erupt, we mean that the force either becomes too strong – and you haven't yet learned what the right amount of force is – so it can erupt according to us. But it can also stop functioning if it becomes too weak. Therefore, it's important that you find a good balance in this.

– Where do we get this molecular iron?

– How did you know we were talking about molecular iron?

– It just popped into my head, so I'm not sure. I don't even know what molecular iron is (we laugh heartily together).

– You're on the right track. It will be a compact molecular iron that is stronger than the one you already have. But this iron cannot be forged. However, it can be assembled.

– What do you mean by assembled?

– The iron, as mentioned, will act as protection for your future motor functions, but it can also be stacked. What we mean is that if we take molecular iron and stack it on top of each other, we get a long chain. This chain is then connected with other molecular iron chains, and eventually, we get a long chain of iron. This assembly becomes much stronger and more durable. It's an iron atom that will create chains of molecular iron and be used as protection.

You will also initiate a new industrial revolution because you will introduce this iron when your industry develops a clean-living environment.

– So we will use it in the industry that produces this iron?

– No, not like that. The iron isn't produced in the industry. It's created in a lab but then developed into a technology that you will use as an atomic substance, in what we discussed earlier.

– Ah, I understand. It will be used as an ingredient where we need to enhance our iron strength?

– Yes, exactly. So the iron ingredient will be created in a lab, where it's then further developed and added to each part where it belongs. If it's for protection, then you add it there. If it's for a molecular iron chain, then you add the technology there, and so on.

But before you get there, you first need to study the sun more than you already do because it's there you will reach even further and deeper into the molecular bonds that connect the iron. Once you've found that, as we mentioned earlier, you can harness this newfound power and complete it before the year 2030.

This iron is based on the inner concept of the sun. So when you've learned and developed the sun's core, your understanding of how iron works will also increase. That's where you will find most of the teachings when it comes to the fundamental concepts that iron can offer you. But to develop it, you need a better understanding of the components that should be included. Iron will shape and influence your daily life in the future, more than it already does. So listen to those who study the sun. And don't forget: it's important that you always share your knowledge with others. We will also be keeping an eye on ensuring this happens.

CHAPTER 12

PURIFICATION - SPLIT

PURIFICATION

Kazandra speaks:

Now that you understand what an atom and the inner workings of electromagnetism are, we can move on to electromagnetism itself. This is an area that needs improvement, which we can achieve by purifying and dividing magnetism. Let's start with purification, and we'll explain how it originated with us.

The purification we're discussing comes from the realm in which we operate – the Kathremic realm. However, the purification we'll teach you comes from a basic perspective. There are more layers of purification that we may write about in future books. But to grasp the effects of purification, we first need to understand the foundational idea behind it. That's what we want to share with you now. Originally, magnetism was essentially a combination of different acting factors, but as we expanded our knowledge, magnetism evolved further.

So, how did we arrive at purified electromagnetism? It happened by chance – at least, that's how it's recorded in our books. But as we know today, nothing happens by chance; everything is created as a precondition for us to develop our teachings, regardless of which planet we originate from. But that's how the masters of the past taught us that purification came to be.

A master discovered it when he tried to set aside magnetism and left it in his warming chamber, where it came into contact with accompanying particles. This resulted in the particles increasing when the magnetism interacted with its surroundings. So all it took was heat and particles to enhance the electromagnetic concept.

– How did the particles get there? I ask Kazandra.

– They were already part of the magnetism we had at the time since they were carriers of a magnetic power source even back then. But as the particles increased, the power transformed, and our electromagnetic ability grew stronger, which was incredible for us even then. However, as with everything else, we can't further develop an inner concept without first having an external influence, which is what the heat provided.

So, the foundation we were taught at the start of our purification process consisted of particles and heat. That's also what we want you to begin with – enhancing the particles' power during the transition from magnetic to electromagnetic ability. You'll soon notice how magnetism can expand, which is the whole point. When magnetism expands, it can expand even more, and this brings us to the concept of layering. You can open up a magnetism just before it transitions to electromagnetism, and that's when you can expand it almost as much as you want.

So start by studying this layering effect to see how magnetism expands, and you'll find that in the next phase, you can continue to develop electromagnetism.

To purify means to improve electromagnetism. So, to achieve a better version of electromagnetism, we must introduce purification during the transition from magnetism to electromagnetism, which you do by assembling a magnetic ability while it's developing into electromagnetic ability.

But before we can introduce a magnetic ability, it must first be able to transmit, or in this case, produce heat. It's crucial that it can do this before we can enhance the type of magnetism we previously mentioned. That's what can then be developed into an electromagnetic ability.

Think of it this way: Electromagnetism is particles that increase when exposed to *combustion*, as we call it. But before we get there, a purification needs to occur between magnetic and electromagnetic abilities. Only after that can electromagnetism maintain its power in amperes, if you like. But we call it a power source, because that's what it really is.

– What is purification? I don't quite understand.

– Purification is when we transfer current between or within a developmental phase. In this case, purification means improving the transition between two phases of development. So purification is a clean and refined concept.

– How do we do that?

– While you're developing a magnetic ability into an electromagnetic ability, you're in an intermediate stage, and it's during this stage that we need to enhance the power before it fully transitions to an electromagnetic ability. That's how you increase electromagnetic particles. But before you get there, you first need to improve magnetism, which is a bit outdated according to us. That's why it's difficult to purify today's magnetism, but that will change.

– How will we know when we have the right magnetism that can be purified?

– Your previous researchers will reach that point. Until then, our advice is to keep developing where you are now and take it step by step. It will come together eventually. Then you'll understand what we mean.

Alfredo continues:

It's also true that we can't always, nor do we need to, purify magnetism. The components you'll need in the future require different magnetic strengths. Therefore, you can't always manage magnetism.

When we talk about breaking magnetism's transformation before it turns into electromagnetism, we're referring to the blocks (plates) of magnetism that you'll also use in the future. So what we're discussing is the part that can be purified, nothing else. These are blocks (plates) that we use on our ship. But for the blocks to function correctly, they need to be purified at least three times. They need to go through the layering effect at least three times for the transformation to become strong enough in its attraction power. But to be able to manage and harness the power in a warming chamber, it also needs *functional improvement*, as we call it.

It's only after you've created these magnetic plates that can attract and transform the sun's power, and when they can be functionally improved and manage their power source, that you'll have achieved the purification we're talking about (we'll discuss how we can manage and functionally improve magnetism later).

To purify the right magnetism in the future, it needs to consist of a material that can carry the sun's power, and this is where we get to the power that your newly developed iron can reach. Only that type of material can be used for such strong magnetism. This means it can be purified enough times to make the attraction power stronger and stronger.

Once you've found the connection between the right material and the correct positioning, we move on to the next part – we convert particles.

– So, it's a purification that's missing for us to achieve even stronger electromagnetism?

– Yes, exactly. But keep in mind: for a purification to work, you also need to add more particles. So, you need to add something to increase something. That's what we mentioned earlier.

– Which particles are those?

– We'll give you a clue: They can't be controlled, but if you increase the number of them, you can change them. The more there are, the more you can increase the power.

– So they're added between magnetism and electromagnetism?

– Yes, exactly. It's also important that you do this, otherwise, the electromagnetism will diminish between each step, between each process, leading to a decrease in electromagnetic strength.

Finally, we'd like to add that there's a *closed opening*, as we call it. When it's no longer possible to purify a magnetic ability, we call it a closed opening because the force required for purification is not always synchronized with all the material. In such a situation, we need a new opening and the right materials. By doing this, you'll soon learn which aspects of magnetism can be purified.

– Is it not possible to purify all aspects, or what do you mean by aspects?

– By aspects, we mean the force needed to increase the power, which we do with the help of particles, the ones we bring together, the ones we can purify. Then we can divide that force so the particles can interact again, and this is where a so-called closed opening occurs. It means that it's closed but can reopen if we add more force so that it can continue its purification.

Summary:

To transfer heat through magnetism, we first need to create it, which we do by using heat-resistant components. Then we can transfer the heat from one side to the other, like a passage, you could say. It's in this current-carrying cable, or link, where we can manage and increase the particles that we introduce into a purification. But to do this, we need something in between where we purify an event, between magnetism and electromagnetism.

HOW A PURIFICATION MAINTAIN ITS POWER

Kazandra speaks:

To ensure that a purification retains its power and internal strength, we first need to stop it. Otherwise, the electromagnetism becomes too strong. And it's by stopping the addition of heat that we can halt it.

But it's not just the heat itself that causes the particles to increase. They also increase through the convergence that occurs. The power actually increases just by them meeting each other, without us adding heat to the particles. They actually consist of a very special force that comes from the magnetic aspect of a star (we'll discuss this in the next chapter). These particles are created specifically in the core of that star. Therefore, their interior is already undergoing an ongoing change. They are active without the need for additional heat, though heat further increases their power.

– Which particles are you referring to? I ask Kazandra.

– They are fundamental concepts of the star, and they are the building blocks of a unique particle we call *the emitters*" – neutrinos.

It is out here in the universe where neutrinos began their first journey. They were created here, in this particular star, long before neutron stars came into the picture. It is a misconception that neutrinos originally come from neutron stars, but that is not the case. Yes, they are created in a neutron star, but their origins are from a completely different star. This is also where our magnetic power gained its momentum, thanks to the effect neutrinos have on magnetic ability.

– So, it's neutrinos we should introduce in the process between magnetism and electromagnetism?

– Yes, but we can also introduce other particles found in a star's *sack*, as we call it. But that's something for a bit further down the line. So start with neutrinos and see where that takes you, and call on us if you can't activate the power that neutrinos can generate through electromagnetic ability.

What more can we say? To ensure that magnetism retains its power and continues to function, it also needs to be maintained. By "maintained," we mean that it can be controlled and retain its strength so that it can eventually be regulated. It's essential that we can control this type of magnetism, or else it might become too strong or too weak. Therefore, we need to maintain the magnetic ability so that the current-carrying force occurs in the right amount and at the right pace.

DIVISION

Kazandra speaks:

To achieve better movement, a better effect, we also need to divide electromagnetism into two parts. We mean the part where electromagnetism has added more particles in the intermediate state and thus increased its power. It is that part of electromagnetism that we will use in the future as a fundamental concept. And to reach an improved version of electromagnetism, we first need to learn to manage the power.

We are not talking about the power that already exists, but we are talking about the power that will exist in the future, where the power can be divided into two parts.

By dividing, we mean, as many already know, splitting into two. But in this particular context, a division implies an improvement. However, we cannot divide electromagnetism until it is fully complete. Therefore, we first need to learn to purify the magnetism. Then, when it has undergone purification and the number of particles has increased, the power also increases. We can then extract the same power again and increase it even more. This is the part we refer to when we talk about division.

– What will we use it for? And why does it need to be divided? I ask Kazandra.

– We divide it so that the power, the one that is created, can replicate itself. This means that after a division, one part will continue to work, and in the meantime, we remove the other part because it is the one that will later replicate its own behavior. We do this so that the concept does not stop. For each process that occurs within the magnetic world, we will always need to divide, because we will always need a backup.

For example: Think about what you do when you copy a piece of paper – a text. This is also a kind of division. You insert a piece of paper into a copier, and then you copy the same text onto a new piece of paper. Same thing here. Because think about what it would do to you if all the information disappeared before you had time to copy the text. It acts as a security arrangement, we can say. Therefore, we need to learn to divide an electromagnetic ability.

– When in the process do we do it?

– You do it after you have started the process of *creating improved elements*. Your researchers know what we mean. We also need to know where it can be divided so that we get all the ingredients in the production so that it can be reproduced. Your professors already know this, those who work within the same concept. But what we want to add here is that the division will come later than you think. You therefore need to wait even longer than you already do when you go in and break and divide an electromagnetism.

That's why you haven't yet managed to extract all parts from a division, and that is also the reason why things are a bit stagnant for you in this matter. So be patient and wait for more time and space so that the internal particles have time to fully transform before you divide it. Sometimes it's not more difficult than that.

POWER DISTRIBUTOR

Alfredo speaks:

We can also make use of a power distributor. For if you ever want to learn how to manage the power that the sun's energy provides, you also need to gain a better understanding of how all the particles behave in the internal motor functions that the sun consists of. And this is where we come to the range of particles in a foundational light concept.

For the sun's power to land correctly and with the internal force that the sun's rays carry, it is important that you first highlight the particles you will manage in your future transformer. Because if you do not achieve the transformation of particles that needs to happen, so that the power increases, you will not get the internal motor functions that the transformer should consist of going. For it is precisely the particles that are important in this, that they transform the right power. Otherwise, you can never achieve the same fantastic results that we do (we will discuss the transformer later).

So how do we understand the scope of the significance that the particles carry so that we can manage them correctly? We use a power distributor. One where the particles themselves can go in and divide, as we previously discussed.

– What does a power distributor look like? And how can it carry power and transform the particles we have in the transformer? I ask Alfredo.

– How did you know we meant a power distributor with carrying power? You are delightful.

– I don't know. It just felt that way.

– Because you are once again on the right track. A power distributor carries power, but it can also distribute power. So each particle can, through a power distributor, itself distribute the correct power in the instrument you are using – in this case, a transformer.

– How is it possible to divide particles? I mean, why can't they just stay as they are without a power distributor?

– Dividing particles has many aspects. Partly, we want to copy them so that other parts can continue along the same path, and as we discussed earlier, so that the power itself does not disappear. We also divide them to then be able to merge them again. Think of it as an unstable relationship. You separate for a while to heal individually. When you then meet again and have healed from the past, you become much stronger together than you were before. That is why we divide particles in electromagnetism so that when they reunite, they can increase their power even more.

This is the part of a division that the power distributor can do for you – where a division occurs but then everything is merged together again.

Your future transformer will also receive another addition, and it is where we add more power. It is a part that comes from the mirage technique we discussed earlier. For it is a mirage technique, within the meaning of the power distributor, that you will need to understand in the future. Therefore, a mirage technique becomes important for better understanding the task of a power distributor. A mirage technique does not constitute an obstacle but is merely a way to achieve more power.

FOUR COMPLEMENTS

Kazandra speaks:

We can also learn to unite electromagnetism, which occurs by adding various *complements*, as we call them. For it is in the actual outfitting of different complements that we can unite a magnetic ability. It is different complements that enhance magnetism so that it can increase its power. We usually say that there are four different complements. These are various sets of events, of a process, that each step must take.

1. Involves increasing the power. We call it *the rising power.*

2. Involves transforming the power. We call it *the transforming process.*

3. Involves transferring power. So in addition to uniting magnetism, we can also transfer power to another. We call this the *transferor.*

4. Involves when or if magnetism diminishes. Then we need to strengthen that power with more segments that the magnetism needs at that time. We call it *a stop.* The magnetism simply stops because it needs new power.

These are our basic complements within magnetism – four different steps where we need to study magnetism in various events. These are four complements that you will also learn about all over the world. So complements are different steps, different directions, that magnetism can take. It can, therefore, choose between increasing its power or staying where it is at that moment. In that case, it does not undergo a new transformation.

It is also within these different complements that you can learn when and how to unite different forces within the magnetic perspective. Otherwise, magnetism becomes outdated. So study these different complements and you will see how you can suddenly unite different segments that that specific magnetism contains. But also how far you can stretch a magnetic ability over time (hint), even to what you do not think is possible today.

This is a universal power, and it can never be controlled by only one source. Therefore, it is important that we can read all the components so that you know where the power lies. Otherwise, you lose connection with the power, and everything can come to a halt. But don't worry, you will learn this. That is why it is now our task to guide you into the era you are about to enter, where we help you rebuild your world.

TEMPLATE

Kazandra continues:

We can also create a template. For us, a template is essentially a dispatch of propagating elements. It sends out and positions itself as a template, even though everything happens internally.

When particles move, they usually follow the same path. This too can be seen as a template, but here we still prefer *an action within the process that takes place*. However, if we manage a force that doesn't always follow the same path, because sometimes it can act a bit disorderly, but we want it to follow a certain path, we can set up a template specifically for that purpose.

– Do you mean we can draw up a template so that the force, the process, that occurs internally is then guided by the template we've drawn up, but only in cases where the particles or the force seem a bit disorganized? I ask Kazandra.

– Yes, exactly, but not entirely. We don't draw up a template for particles that cannot be controlled. We want them to act a bit irregularly at times. We create and apply a template so they can behave differently and develop into the force we want them to become.

– Doesn't that involve a transformation then?

– No, in this case, we don't transform. We teach the particles how they themselves can follow a certain path, based on a specific template. That way, they can strengthen their own power. There are particles that sometimes behave irregularly, which is their natural process because the force is so strong.

But sometimes we don't want them to behave irregularly. Sometimes we want them to move straight ahead, and this is where the template, the new path, comes into play. If we can get particles to move in the same direction, they eventually perform a kind of self-transformation. And as more particles do the same thing, we also build a stronger supporting concept. Therefore, magnetism becomes more stable if we guide the particles to move in the same direction, meaning they cooperate. When they do that, the force increases instead of being disorganized.

– So, we redirect them so that they can learn to take another path, and the more there are, the more the force increases when they follow one and the same source. Is that what you mean?

– Yes, because that's what we want. But there are also particles that already follow one another, while some act a bit irregularly. And it's only that part of a magnetic process that we want to change, nothing else.

You will also integrate the template into a computer program. There, you can add the power that your magnetism is capable of carrying, and where you introduce another section in the program for those that are a bit more irregular. By doing this, you can utilize all the power that exists in a particle's capability, not just when they already act as we want (we will discuss the computer program further later).

Think of it as a product where you maximize your use of that product, instead of discarding the part that no longer works as you want. It means that you use the properties of an element to its full potential and don't stop just because you think you can't develop the product further. Because it's always possible to revive or change a behavioral pattern in a particle, as in this case, so that we can use them for other purposes as well. Of course, there is a limit, but that limit is far from being reached when it comes to your magnetic capability.

OUTDATED INSTRUMENTS

Archangel Metatron speaks:

There are also quite a few measuring instruments that need to be improved in this regard, as we believe they are outdated. Which instruments are we talking about? The section where the magnetic range begins is the part that has become outdated. There are some core instruments within that section that are outdated, resulting in inaccurate measurements.

How do we know this? We know because the results you've reached are, in our opinion, not entirely correct. They're not completely accurate in their final outcomes. So it's not that you're measuring incorrectly; it's that you need to improve your instruments. That's why the final results concerning the magnetic range aren't entirely accurate in their calculations.

So how do we achieve the correct range that reflects the right results? You do this, as we mentioned earlier, by starting over from the beginning regarding electromagnetic capability. That's where it has been distorted, and where the range of internal power and capacity has failed. Once you've done that, you can improve, not create new, the instruments you're using today.

In the future, a completely new technology will be introduced, created by English-speaking souls, and they will lead that part. It will be a complement to the tools you already use, thereby allowing you to improve today's instruments. The complement will also exert its power in amperes. Therefore, you need an instrument that can expand and increase the circumference in an electromagnetic ampere, if we can call it that. Because it is here, when we expand and increase the power, that the range of power has become misleading.

So, seek out a complementary part that is a better fit and that expands and increases electromagnetic capability. Then add the power you obtain and multiply the range further with the power that provides a better and more accurate result. In this way, you'll achieve a better range of the power that an electromagnetic capability can express. In other words, it's about the circumference that surrounds magnetic movement capability, if you understand?

– I don't fully understand. Could you elaborate? I ask Metatron.

– Of course.

To calculate the range of magnetism, we need tools that can measure the power it consists of. It needs to capture the power to measure it. Once you've captured the power, you can measure the range of power that an electromagnetic capability emits.

– Is it only the electromagnetic power that emits a range of power, and is that the only one you want us to calculate?

– We want you to retain the magnetic capability but develop the electromagnetic capability. So both can be calculated within what captures the power. But it's the circumference (range) of electromagnetism that we want you to focus on.

– How do we calculate the range?

– It's done almost the same way as when you calculate the circumference of a circle. Sometimes it's no more difficult than that. And it's a technique you're already using. You just don't have the right equipment to achieve the results we want you to reach. That's why you first need to calculate the range within the reality you live in. But to achieve further results, you need to supplement with new methods, new instruments, and charts that can explore the electromagnetic range of power. Only then will you reach the correct results with the right power.

CHAPTER 13

THE MAGNETIC POWER OF THE FUTURE

MAGNETIC BUBBLES

Kazandra speaks:

When you're ready to further expand your knowledge, you can also make use of the magnetic bubbles that surround your world. This is where you can gain more insight, both in terms of electromagnetism and how it affects your magnetic field. These bubbles were only recently discovered by humanity.

– How can the bubbles affect our magnetic field? I ask Kazandra.

– It doesn't affect it in the way you might think. At first, humans will believe that both work together, but the bubbles aren't controlled by the force that normally influences a magnetic field. However, they are affected by the magnetism present in the middle atmosphere, and that's one reason we're telling you about this. This is where you'll further develop the study of magnetic fields, especially in terms of how a magnetic field can be divided into different forces, each having different effects depending on where in the universe the field is located.

Once you understand this part, you can continue to expand your knowledge of electromagnetism.

This is a field that has been with us for many years. Therefore, we now want to tell you about its origins and how you can improve your magnetic abilities with the help of these bubbles. But to gain knowledge about these gigantic bubbles, we first need to travel there. So, I ask you, Ulrika, to take an astral journey with me.

I get up and move to my meditation corner. I light a candle, sit down, and begin to breathe deeply and calmly. Together, we travel into the world of the magnetic bubbles.

– We're now going to take you into a small part of the world of bubbles. So don't be afraid, my friend. Just describe in your own words what you see, Kazandra says.

– Thank you, I'll try.

I begin my journey, and after a while, I find myself in the universe, approaching these gigantic bubbles.

I see enormous bubbles, arranged in two rows, maybe more, with an acting magnetism in between that holds everything together, almost like gravity does but not quite. There is a bubbling content here, but it's a more subdued bubbling, like sulfur or sulfuric acid, but it's not sulfuric acid. It's something else that resembles it. This is a new and somewhat dangerous substance for us humans, and it should be taken very seriously by the scientific elite working in this field.

I also get the sense that they can't be controlled, which is why they're so difficult to penetrate. The outer shell is incredibly strong yet soft, like plastic. A strange feeling, but a natural structure for the universe.

– What else do you see, and what do you feel?

– I feel a sense of peace; they give off a calm. Without these walls of bubbles, anyone could enter our galaxy from other parts of the universe. But I also feel some concern. There's a weak link here, and that's what worries me.

– Explain.

– I sense there's a passage between the walls. A free passage, where something or someone has managed to get through but has remained between the walls because the magnetism affected them negatively. The more controlling part can then lose its power, and the bubbles become more permeable. Is that correct? I don't fully understand.

– Yes, almost. You're right that there's a passage between the walls of bubbles, but there's also a barrier that sends out a warning if someone tries to break through. We, the Galactic's, can then quickly get there and help with what's happening. So, you're correct; there's a hidden door and a means to destroy the bubbles. But we are strictly forbidden from giving you information about this, both to protect you and our world. What else do you see? Your vision is wonderfully penetrating – continue.

– I see large halls, but they're not like our halls on Earth. They are vast rooms where we can pass through if we want to gain more knowledge about past and future research. I don't fully understand, but the bubbles can teach us more about how to travel between time and space, through different past, present, and future halls.

This is also, I believe, where the halls come in, in different dimensions, to pass through to move forward. This will also help us better understand Einstein's theory of relativity. But to do this, we need to be spiritually developed; otherwise, we'll only see giant bubbles and miss the rooms that we can only see within our minds. But this is also a future opportunity for us humans to grow, for those who wish to.

– Exactly. But it won't be an easy journey to understand the significance, protection, and content of the bubbles. However, humans will travel there in about 10 years, and you will encounter these passages between the walls and take samples and images of what's there. However, you'll need to wait to explore the contents, for safety reasons and as long as there are egocentric souls on your planet. And you're right that this is where we learn to bend time and space. Tell me more. Where are you now?

– I'm inside a bubble. It's ejecting me. An explanation follows: The bubble's contents can build up significant pressure that it occasionally needs to release, much like a volcanic eruption. Otherwise, the walls and contents could collapse. So the bubbles need to release a little now and then. But what happens to the contents that are ejected into the universe?

– Not much happens. The contents the bubbles release doesn't have a direct negative impact; everything is taken care of, partly by the magnetic ability that already exists in between. But there's a limit to how much each bubble can contain, and as they activate new substances, they need to release the old ones to avoid their own explosion. You see, sometimes a bit of tension needs to be released to maintain future calm.

– Do bubbles ever explode?

– It happens, but neighboring bubbles quickly seal it off, so the chain never completely breaks. They are self-regulating, a self-sustaining process.

– Thank you, my dear ones. I need to rest now. It's affected my soul, but it's been a fantastic journey.

– Thank you too. We'll rest and resume our contact at a later time.

I come out of my astral journey, and while we were talking, I had already written everything down. For its only with my outer eyes that I write, but it's within me that I see what's happening in the universe. That's usually how I work – double.

After a long rest, we continue our conversation in the kitchen.

– What do the bubbles contain, and why are you telling us about them in our book?

– They are magnetic bubbles, and just as they sound, they are bubbles. They are large and act as a barrier between certain galaxies, both as protection and as a concept for the universe to produce. For they also produce, but that's a process we won't go through now as it's too early to share that information. They consist of a material and substance that doesn't yet exist on your planet. That's why we only mention it for future days. But we can say this: To harness the right electricity in the future and bring home electricity from these magnetic bubbles, you can use the outer layer of the bubbles.

It has come to our attention that you aren't yet handling electricity in the right way. This is because you haven't yet developed the technology available within the bubbles. When you reach that point and study the outer shell of the bubbles, you'll understand what we mean. For it's the outer shell that acts as an active power source, a current carrier, and that's what you need to learn to activate.

So find a good solution to continue observing the bubbles. Consider using a current detector, and you'll understand what we mean. You will then encounter a different type of ampere and learn how to take advantage of this fairly simple technique to generate power. Once you've learned about the outer application, you can move on to the inner.

It's in the outer where the power source is found,
and the inner is an enhancer of an
electromagnetic sphere.

These outer bubbles have also caused many conflicts in the universe due to their vast number. So the power, the inner capacity they possess, is enormous, which you will soon realize. And it's this power that some Galactic's have learned to use in a way that is very wrong for us.

– How can we use this technology on Earth?

– If you learn to apply the same technology that exists there, you can use it as a powerful shield. That's why they surround many galaxies in the universe as protection. But listen up! The universe is a shared community where we all reside, and we only serve its good purposes. Therefore, we can't reveal more than this, but you will learn from the bubbles as a way to protect yourselves, including against the strong gravitational forces in the universe.

The bubbles will also enhance your knowledge of the behavior of magnetic fields and electromagnetism. That's why it's important that you continue to research in this area. Through this, you will arrive at the level of electromagnetism that we use on our ship. But the most important lesson in magnetism is the knowledge of how to conserve and manage energy. For what good is power if we can't learn to manage it?

THE MAGNETIC ASPECT OF THE STAR

Kazandra speaks:

You can also tap into the star's magnetic aspect. We've named it *the star's magnetic aspect* based on the teachings from our planet when we developed this concept, which we are now going to share with you. This aspect forms the foundation of what will soon come to you as well. It's a magnetic force that originated where the star Sirius is located.

We know that Sirius and the surrounding stars are already well-known to you. But only we understand why this force hasn't been revealed to you yet. Due to the times you are currently facing, we've collectively decided to briefly share the information that was given to us when we entered our magnetic era.

Sirius originates from another time and another part of the universe – a time when your solar system did not yet exist. That's how ancient the technology we are about to describe is. It's a technique, a teaching, that was created back then, which some galactic beings picked up on and further developed among us.

The star's magnetic aspect is about how we can manage magnetism over a long period, as that's where sustainability is found. It's an ongoing process where everything is transformed into different sections. It actually divides everything into sections on its own, with each section having its own power and strength. This is how it achieves sustainability. Think of it as a star that expands and contracts – contracts and expands, and it keeps doing this continuously. So, *manage, sustain,* and *transform* are three key concepts within the star's magnetic aspect.

– Doesn't it eventually explode like other stars? I ask Kazandra.

– Yes, it does, but it takes much longer for that type of star because it's more sustainable and long-lasting. By dividing itself into different sections, one part can slowly decrease in activity, while another part rests. Then, another section takes over and increases the power, and after that, another part takes over again, allowing it to slow down once more. Then, it all starts over again. This is how it develops its internal sustainability.

This is the part of your process that you need to develop. If you do, your magnetism will also keep going, just like the star. But to arrange a magnetic process where it moves in and out – out and in – in a constant motion, we need remote control that can manage and direct where we want the magnetism to move (we'll discuss the remote control later).

There's also a space that can store magnetism, and it's this very space that is in control. Without it, you wouldn't be able to restart your star-magnetic capability, as it acts like a *core*, which we call the core house. That's where all the extra storage happens. Think of it as a reserve tank. If our ship or something else were to break down, or if something happened to our magnetic ability, we wouldn't be able to travel as far as we can. That's why it's important that you first learn this principle before anything else can take off.

AN INTEGRATED – MAGNETIC APPROACH

Kazandra speaks:

How, then, do we integrate the star's technology with the reserve tank that we need? We do this through the knowledge of an integrated magnetic approach. This is merely a method and cannot be created.

So, what does an integrated magnetic approach entail?

- An integrated approach is based on what we can bring together.

- A magnetic approach is about managing the energy that we have gathered.

Once you've learned this, you can use your magnetic ability for a longer period. But when we manage a magnetic approach, there's also a risk that the magnetic ability will lose its power. And if it loses its power, its approach, it can negatively affect other things as well.

– Why does it lose its power? I ask Kazandra.

– Because it is the controlling part. Think of it like a driver losing strength. It doesn't really matter how powerful the engine is then, does it?

– True.

– So, what you need to do to prevent your magnetism from losing its power in a coupling system is to maintain a spare part. One that can always be repaired, one that can step in and read the situation when needed, one that never loses its power. Think of it as changing the driver every hour.

– How do we do that?

– You do it with an extra battery, we might call it. But here we're talking about an extra magnetic source. This source will control your magnetic ability if the original one loses its power.

– I still don't understand. How can we connect it with our coupling system? Where would it be located?

– It's always important to have an extra option. But what's even more important is where it's located because if we place it somewhere it can't step in and connect itself, it won't work. So it's crucial that your engineers find the right location for this extra power source, or else it will fail. Once you've found where to place it, you also need to integrate it with other components that will be part of a remote-control system so that you can monitor it.

But we'll give you a clue: *It can't be seen, but it mirrors the power that can be used.*

Once you've accessed this extra energy storage capability, it never runs out because when the original rests, another part takes over. This is what we mentioned earlier. But as soon as it steps in and takes over again, it opens up a new power, a new channel, that can continue to manage energy. That's why it's essential to have different channels because each channel contains different propagating elements.

By propagating elements, we mean those that don't pose an obstacle, those that spread and can step in whenever needed. They never stop; hence they propagate their elements – they guide them into the order they belong in. In this way, they also function as a controlling element. We are not allowed to elaborate more on this.

By storing magnetism, we can draw out pure energy whenever needed. It's like storing gasoline in a gas can, but instead, we store plates filled with magnetism. These are what we discussed earlier and will discuss in more detail later. Additionally, you'll add a movement component as an integrated part, including a time unit, which we'll discuss further down the road. That's why it's important that you adopt the right movement for the right unit, or else it won't work.

– So there will be different units for different magnetic movements?"

– Yes, but not exactly. There will be different units for different magnetic abilities, as you mentioned, but that's further into the future. And as you know, we always work from a foundational perspective. So, at first, you'll only have one unit that provides a magnetic movement ability, and that's the one you'll use first to learn how to store energy. And to replenish energy, you'll use the power of the sun. It's also the sun's magnetic union that will teach you how to connect both.

The Sun's Magnetic Union wishes to step in:

We wish to share our part in this work and give you the knowledge we will release in the future when the time is right, which is the knowledge of the sun's magnetic power. This is what you will use as a storage source in the future, like a fueling station, you could say. That's why we've already taught you the knowledge you need about solar cells so that in the future, you can use them in all areas of magnetism.

Once you've reached that point, we will give you new tools, for example, through Ulrika. Then all the old information will be erased. It's essential that you start fresh on this matter because there's also a deliberate, built-in stop to prevent your development. Once you've learned to store and manage fresh, clean energy, your development will progress quickly. Listen to those who speak of new things, not old ones. The old only tries to mislead you back to where you once were.

If we can't store it,
we can't use it neither within magnetism.

Kazandra continues:

Once you've learned to manage energy, you can move on to

other tasks. What you need to do then is to look for a motor function. In this context, by motor function, we mean the part that will work with a magnetic ability. It's like your current car engine, but in this case, a much smaller part. It will belong, but it does not constitute the main part.

– I don't quite understand. What will work with what?

– What we mean is that the elements you choose to use determine what kind of motor function we're talking about. If you use one type of motor function, you need to adapt to it so that the elements you want to add are also compatible with other added components. This is what you will then be able to connect with the sun and magnetism, everything that provides movement and everything that can store energy.

– What will the motor function consist of?

– It will consist of neutrinos. These will provide the speed, like your engine does today with an associated car battery. That's why humans have already studied the power and content of neutrinos. And yes, it is possible to both store and manage neutrinos in magnetism.

But to store kinetic energy, we first need to build a container where everything can be gathered. This is what we mentioned earlier when we talked about units and generator blocks. In the beginning, your generator blocks will use magma to produce heat. But later on, you will also use magnetism as a micro part to store energy using both neutrinos and protons as a control force. That's why we mentioned earlier that neutrinos and protons are an independent force.

The magnetic approach is about managing energy, but by this, we also mean transporting energy from one part to another. For we believe that this, too, is a way to manage energy. It doesn't just stay in a box; through the movement that occurs when it's transported, it remains and increases, continuing to release its power.

When we manage a magnetic approach, it's fundamental that we can also conserve the magnetic ability so that it doesn't diminish. This is where the eternal cycle comes in. For if you can maintain a power that just keeps going, you don't need to transform it anew, because it's always there, being transported from A to Z.

That's why it's important, in a magnetic approach, to also see the transport path itself as a way to manage energy. Because when we transport energy, we can also carry it more effectively. In this context, we mean that it's always strong, that it always has the strength to uphold the power it was created for.

Once you've understood this carrying, transporting phenomenon, you'll move on to the next part, where we begin to transform particles to increase the effect. And the more we increase, the faster the transport path becomes, and the more we manage our magnetic energy potential.

So you see, without knowledge of *how to manage energy in different ways*, we can't fully use the magnetic ability. Without it, we can't transport the power in our conductive source.

MANAGING ELECTROMAGNETIC ENERGY

Kazandra speaks:

As we previously mentioned, it is the electromagnetic capability that you need to focus on and further develop the concept to increase your energy. This is why we talk about electromagnetism as a concept for the future. It's also something that has been distorted to teach you incorrectly.

So, how can electromagnetism be distorted? Besides what we mentioned earlier about the incorrect composition of baking powder, it is done by assigning others an incorrect magnetic background. It's the circumference of the magnetism that has been distorted to prevent you from fully developing that aspect. Because if the circumference of magnetism cannot complete a full rotation, you cannot correctly convert it into proper electromagnetism. You think you've done it because your electromagnetism works so well, but the part that isn't functioning is the one that has been blocked.

– What is the true magnetic circumference? I ask Kazandra.

– The circumference we're talking about can actually be multiplied and converted more times than you're aware of. That's why you think you've reached the peak, but you haven't. In this case, the circumference can find its own path, where the energy keeps flowing, which leads us to the circumference that the electromagnetic capability will be able to manage in the future. And this is where we come to an integrated magnetic approach. Because in this context, it's electromagnetism that we'll manage, not magnetism.

– What's the difference between managing the two?

– Both can manage a *release*, as we call it, and the energy that is released can be stored and used later.

When it comes to electromagnetic capability, we can use the same energy over and over again without having to manage it the same way we do with solar energy. So, there's a significant difference between how we can use magnetism in the future compared to today. That's why not everyone understands our words today, but you will in the future.

The electromagnetism that will be managed in the future, the one that just keeps going and going, is the one we will store and further develop so it can synchronize with other objects. And it's crucial that we don't just stop development there, but continue to develop the negotiable, newly emerging forces that you've begun to develop.

But before you fully understand how to manage electromagnetism, you first need to learn to stop it because we can't manage a force that just keeps going unless we can first capture it. So, we stop it, but only to learn how we can manage it.

Once you've learned how to stop an electromagnetic process, you'll get a new meter that can measure the strength of the force at a specific time. This will be important because we can't manage a force if it's too weak. If it's too weak, we can neither measure nor use it.

– Where does the energy go that we're managing?

– You store it like in a car battery.

– So, it goes into some kind of box?

– Yes, but here we're always talking about different strings because that's the only way we can manage electromagnetism.

– What are strings?

– Strings are just the name for an elongated compartment where the energy is stored. Think of a car battery, if it had different elongated inlets and outlets where you could transfer and store energy.

– Why strings?

– We need strings to be able to read the energy we've managed because each string has its own strength, such as in amperes. But in this case, we mean a self-sustaining force, what you call electricity.

– How can we retrieve the electricity again?

– You do that through a kind of transportation path, and this is where your electrospheric cables come in. Because they will transport the electromagnetic energy as they are much more stable than today's cables (we'll discuss these cables later).

– So, we manage electromagnetism in some kind of container, connected to electrospheric cables? Sounds like something we already do, but with updated instruments and cables.

– Exactly. Why complicate things more than necessary?

But the most important thing isn't just how we manage the energy, but also with what and how we can then open, transport, and put the energy back into motion, which you do by integrating everything into your newly developed software. This software informs you of when and how much the energy begins to decrease. You can then check how much energy has been consumed and how much remains. Then, you connect the electricity to a data system that reaches your cables, which then supply the transformer with more energy. And so it continues.

TABLE

Archangel Metatron speaks:

To measure and manage power correctly, you also need to develop a new table. This will be a table that specifically addresses the magnetic concept. It's a table that many of our galactic friends already use, and you will too, my friends.

The table is designed to be filled with components that are part of our future input. It shows what we need to create, what we need to input, to generate a new power. Therefore, the table also contains only parts of the power, as each particle contributes its own strength and the power it then settles into once everything is complete.

The table will have five columns, where each column indicates where each particle should be placed and how much it increases with each placement. By "placement," we mean the position. We use the term placement because the particle undergoes a placement, and for each column, the particle moves forward as the power increases.

This is our definition of placement: a column where the particle resides in its own strength, increasing its power before moving on to the next column. There, the particle again increases its strength and then moves to the next column. And so it continues. Essentially, the table is created to show how much each particle resides in its power – how it increases its strength with each column, and so on.

So, it will not be a mathematical table. It will be a power table that measures different strengths without you having to calculate the power yourself; it does that automatically. We call it *the transforming power table* where particles line up in rows. But for simplicity, you may refer to it as *the magnetic table*, or whatever name you choose.

It will be an automatic table and not something you can work out on paper. It will be connected to a self-regulating system and help the magnetic capability to determine how much power (transformation) each particle has achieved in that particular column. So, create this table. Once you have transformed the correct magnetic power, input these different moments into the strength.

Archangel Metatron concludes:

These are just a few of the points you need to cover now before you build a new transformer. What good is the power in the transformer and how it passes through each component if we don't first understand how to manage and calculate its strength? Once you've done that, you will enter the next phase, where the transformer comes into play – the one powered by a solar-magnetic concept. Therefore, it's important that you first study energy waves on a deeper level and how particles can increase their power.

CHAPTER 14

TRANSFORMER

TRANSFORMER

Kazandra speaks:

Now that we understand why today's magnetism is not sustainable for the future, because it has been distorted, and we understand how to purify magnetism, we can begin building our transformer.

What is then a Transformer?

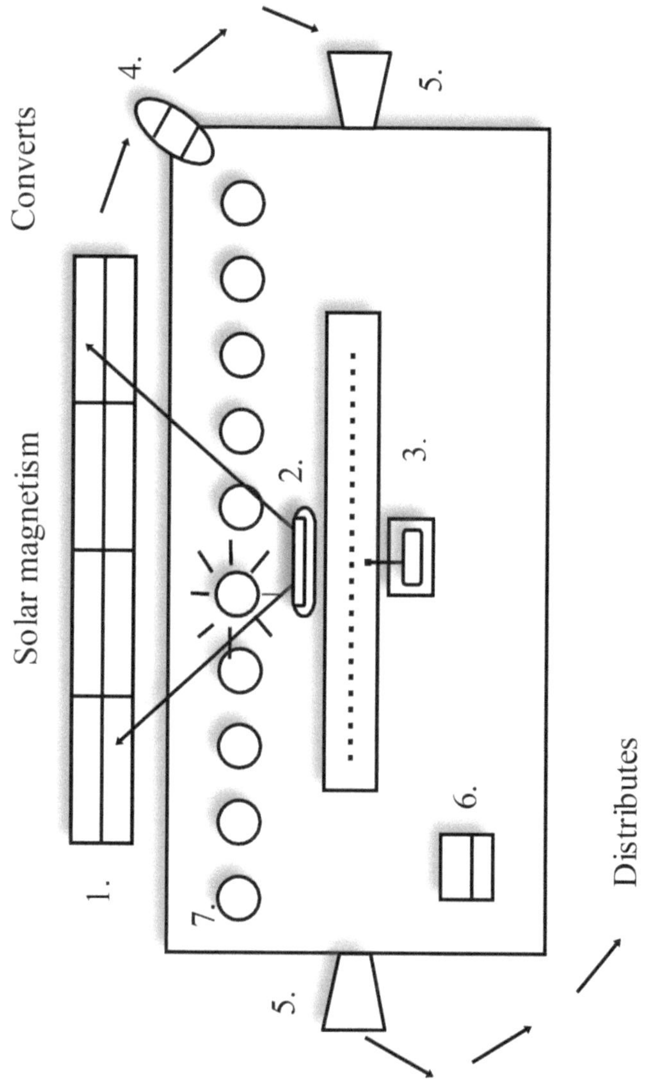
Solar magnetism
Converts
Distributes
1.
2.
3.
4.
5.
5.
6.
7.

1. Solar magnetism.
2. Temperature Carrier.
3. Power distributor.
4. Time capsule.
5. Positive and negative poles.
6. How much power is left.
7. Where and how much the power loads on the inside.

A transformer is a device that provides us with a power that can operate independently. It functions in a way that involves both input and output. To generate heat in a transformer, we also need to introduce something that can increase its own power, which is where the converter comes in and distributes the power that has been generated (see the image).

Thus, it will be an eternal module, not a circuit, because a module operates independently; it can stand on its own. A circuit always operates together with something else. Therefore, we need to introduce a new type of transformer that can produce and increase its own power.

- Therefore, solar magnetism will contribute to the transformation to future electromagnetism and generate heat so that particles can be converted and distribute their full power.

- Thus, we first need to learn how to convert magnetism into electromagnetism in time and space, and how we can store and manage it for future use (we will discuss time and space in the next section).

SOLAR MAGNETISM

Alfredo speaks:

So, it will be solar magnetism that will power the transformer, and everything you will learn from the sun. However, you will add various elements to the transformer through an electromagnetic capability. The electromagnetism you will use in the future is also much more durable and stronger. Therefore, you need a power that can be managed in the same way.

How do we do that? First, you need to learn how to manage energy, including the inner concept of the sun, which we have already discussed. It is fundamental that you understand this before you introduce a new solar power, the one you will create in the future. This is also where your solar stations come into play.

– What is solar magnetism? I ask Alfredo.

– For us, solar magnetism is a plate that captures the sun's power and generates energy waves that are then produced as heat, either to increase or decrease the power (energy waves).

– How can we use solar magnetism to lower energy waves?

– You do this by dividing the particles of solar magnetism into different stations. By doing so, we can turn off a section – that's how we lower it.

– Does this ever need to be done?

– It does happen, but usually, everything takes care of itself in the process we are in. You will eventually get to that point too, and then you will understand what we mean.

– Can solar magnetism diminish?

– It can, but if you always carry an extra set of plates, which we always do, you can learn to replace the power and keep going. But for you, it's a long way off, but it will come.

– Can the sun's power be used for other purposes that we don't know yet?

– Good question because you will do that. The part of the sun that you will maximize in the future comes from the sun's inner core. It is there that you will add your iron power. Does that sound strange?

– Yes, a bit. How do you mean?

– Since almost all iron melts at a certain temperature, it is precisely iron that you need for solar magnetism. This allows you to charge the magnetism when the iron is hot, and that's when we take the opportunity to transform the particles into something else. This is also where the saying "Strike while the iron is hot" comes from.

– So we will use iron together with solar magnetism in the future?

– Both yes and no. Initially, it will only be used in its weight to manage heat, for example, plates, like in a heat cabinet. It is only that part of the magnetic capability that you will connect your iron with the sun.

– So the iron becomes hot, almost melts but doesn't, to keep the power going?

– Yes, exactly. But it is not the iron that increases the power in a transformer; that's the role of the neutrinos. Iron is only used to maintain the heat because, like everything else, the heat will eventually run out. Therefore, you will need, just like us, extra plates that are already filled to the brim with heat-insulating powers, which can be recharged when you bring them to your future solar station. That's why we have a solar workshop on our ship. But to do that, you will also need to learn how to charge particles, which Kazandra will explain now.

Kazandra explains:

To charge particles, we need to add heat, and this is where the solar magnetic capability comes in. It will be essential that our newly manufactured transformer has an entry where the sun's actions are allowed to enter. Through this entry, you will attract together with neutrinos. They will assist you in this process as they are transformed by the sun's magnetic capability.

– I don't quite understand. How do we get solar magnetism into the transformer, get the power in I mean? I ask Kazandra.

– We let solar power in through a kind of ampere that has already utilized the sun's ability and transformed it into magnetic energy. So ampere becomes important in this context and should be connected with the current-carrying link that passes through the entire transformer. This is the only way you can use solar magnetism by transforming the current to reach the neutrinos so that they can, in turn, increase the power.

Normally, we need a force that pulls everything together for us to use the sun's magnetic ability, for example, through a magnetic field. But in this case, we do not need to pull together, just add enough solar power for the neutrinos to perform their task. The power causes the neutrinos to increase in number, and then the transformation happens on its own. Neutrinos pass through a current-carrying link where they operate in their own field.

When selecting a magnetic capability, it is important that it is and remains self-sustaining. This way, we do not need to control or manage the power ourselves as it is transformed. Thus, the sun's power will only constitute a small part, and it will then be the one that heats the particles.

– So the heat always needs to be present, right?

– Yes, it does, hence the ampere acting as a heat source for the transformer.

– How can we create a solar magnetic capability? I mean, how do the particles capture the heat?

– They do so along the way. We will also install a minimal combustion system. This will be what keeps the ampere going. It acts out the power or distributes it from the outside to the inside. It works by capturing particles and sending them onward, which is why it is called a combustor. As long as the magnetic capability is supplied with solar energy, the transformer will never run out of ampere. A combustion system is always made up of its own power and can never manage more than one power at a time. Otherwise, a fault in the process could occur.

It can also act as a heat enhancer. But then it works backward, which we will not go into here; we don't want to complicate things for you. We know your professors will understand this, but until then, you will just have to trust our words … or not (we will delve into the combustion system later).

So, to capture the particles circulating in the universe, we need a transformer that can do that. Therefore, the transformer must be designed in a way that it can act, direct, and harness the powerful energy of the sun.

– How do we create such a transformer?

– Your professors will handle that, and it will be done in the same way that you currently harness pure solar energy with today's solar panels. However, here we need to transform particles that are part of the system we have built. Is it complicated?

– Yes, a little.

We will clarify:

When you build the transformer, it is important that it not only converts energy but also can draw energy from the universe. This will be part of your solar calculation that your professors will work on. They will embrace the concept of *pattern adaptation*, which we and the sun's mathematical union discussed earlier regarding a solar pattern.

Each current-carrying cell, which is what we call each square, can independently calculate how much power that particular cell, that particular square, needs to draw in. Otherwise, the total sum will be either too strong or too weak, and that's what we mean by too little or too much.

– Does each cell or square operate independently in its own work?

– Yes, exactly, and this is where the temperature carrier comes in, as it will determine how much power needs to be allowed in. The temperature carrier will therefore be crucial so that the transformer can achieve an open balance in the power you draw in.

– Are there cells that do not act at that moment, or do they always cooperate?

– They always act independently, even if we have combined them. It is the transformer with its temperature carrier that knows how much, when, and from which cell the power can be drawn. This way, the energy never runs out. When one cell is active (working), another rests and is replenished with new power, just like a star. And so it continues. Therefore, it is essential that they act independently.

Regarding your connection with solar magnetism, your professors will resolve this. But let us say this: Do not take in all the solar power at once. Let them get used to each other, and do not expect an explosive start. Everything takes time to develop, especially when we then need to connect everything. And to do this, you need to learn to calculate anew and develop a new temperature carrier. This is also what the sun's magnetic union will discuss now.

TEMPERATURE CARRIERS

The Sun's Magnetic Union speaks:

In order to calculate *how the sun's rays strike the surface of an object*, which you want to convert the sun's power to, you need a temperature carrier. This concept is part of the magnetism you will use in the future. It will also look entirely different than it does today.

A temperature carrier is a new type of current carrier. It can sense on its own when there is enough power to be converted and used – as opposed to when there isn't.

– How, and for what purpose, can we use the temperature carrier? I ask the Magnetic Union of the Sun.

– It will be used as a measuring stick, for example, in a car. It will be placed in the center of the transformer, where it can sense the incoming power. This way, you'll have better control over whether the power you need to convert can actually be converted. Because, as Kazandra just mentioned, if the power is too weak, it can't be converted. If it's too strong, the power increases too much, making it difficult to capture when and where the sun's rays can be harnessed. This is what your future temperature carrier will detect, making it a necessary component in the future as it can independently calculate and measure its power.

– How does it do that?

– It does this through an integrated coupling system, because everything, when you're done, will work together, and the temperature carrier will then send out signals on how strong or weak the power is – both going in and out (we will discuss an integrated coupling system later).

– How is the temperature carrier constructed?

– Your professors will create it, and it's already in a computer program in the western part of your world. But we can't open it as long as "they" are still around.

– So the temperature carrier is a fairly large creation, if I may call it that, since they don't want us to access it?

– Yes, exactly. Because it doesn't just generate and indicate what power has been converted. It can also illuminate when and at what temperature the best conversion can take place. So it will have many functions in the future, including integration into various applications.

– So we have a temperature carrier that can sense the temperature and the strength of the power being converted. But what happens if something goes wrong?

– If it shows an error, we'll add an integrated pattern of change. It might sound strange, but that's how everything will be laid out in the future, in patterns, to identify where in an event things go wrong. This will make it easier for you to find an error and quickly fix it. Therefore, you will secure everything in a pattern-based process where each change is assessed as either good or bad, and everything will be played out on your data.

Finally, we want to add that regarding the input and output of the sun's mathematical concepts, as we discussed earlier, it will change its amperage, and when it changes its nature, the power does too. You will no longer be able to multiply in the old way because the new way, which you will learn, is more about how we can expand the power so that it just keeps going – becoming more self-sustaining.

Then it's more about the strength that comes in, like what your solar cells encounter, and the power they can't capture. And this is what the temperature carrier senses – whether it's too strong or too weak. So, to be able to add power in the future, you'll need to use an entirely new calculation system. It's a system that's more about building up rather than adding on.

By building up, we mean creating power that can be used and applied in other contexts as well. By building up this power, you can then continue to develop it further. In this way, the power in the transformer also increases. So, it's not possible to use old behavior patterns or old calculations in a future version. You need to change this to fully understand the power that the sun sends out.

There will also be a completely new table in the future that only shows the sun's power range, and this is when you break everything down into patterns. Because, as mentioned, it's the power range that we need to calculate to reach the pattern, meaning the rest of the sun's external capacity. This is because the transformation that happens when the transformer reaches the sun's power can never be converted on the outside. The power controls the outside, yes, but the actual conversion always happens with the help of neutrinos on the inside. They transform when the power increases, as the number of neutrinos increases.

HELIUM

The angels speak:

We can also use helium in our future solar project. Helium is a fundamental element that arises through a fusion process in a star. And to be able to use the same helium found in a star, we need to act in the same way a star does.

So, we need a container and something that can create and transform our content into helium, which we know you already can and do. But how far are you willing to go with this? How far are you willing to use your helium to produce an entirely new substance? A substance you've never tried to create before. That's how we need to think now – to use what we already have and transform it into something new, even though we know some people might be skeptical about it.

– Why helium specifically? I ask the angels.

– Helium is a far-reaching substance, which is why it travels far. And because it can travel far, we can also extend the chain much further than science can and does today. But this can only happen if the concept is combined with other far-reaching components, in the same way it occurs and is extracted in the universe. For it is indeed extraction that we are after, because we know that helium at the right temperature and under the right conditions can be extracted and expanded, making it extremely far-reaching.

This is what we want to convey, because it is precisely helium that you will use in the future, even in certain machines. Therefore, the Kathremes will now explain how they manage helium on their ship, together with water, as a far-reaching substance.

Kazandra explains:

We will try to explain how our process works, the one we have on our planet and our ships, and how we use the power of the star to act as a current-carrying link.

Not all stars contain something that resembles your water. For us, it has a completely different meaning, and the formula we use consists of a D. D symbolizes the amount of water we need to act as an enhancer. It's similar to how you produce hydrogen, but to achieve the best effect, we add helium instead.

We have divided it into three sections:

1. Imagine two containers. Similar to your glass bottles with a long neck but a large body. In one, there is water, D, and in the other, there is helium. At the bottom of each container is a magnetic plate, which we call a block, that pulls down what we need for our transformation. These are the blocks we discussed earlier. The transformation occurs just like in a star. Everything is produced in the core, which the outer layer protects while the process continues.

2. In between, we use a blue light that generates heat, which speeds up the process. If the conversion of particles increases, the speed also increases. As a result, steam is generated, which functions as a current-carrying link, providing power. Then it moves on to process three.

3. Process three is powered by the sun. This is where we charge our magnetic blocks before departure. The magnetic blocks direct us in the direction we are to travel,

> almost like your navigator but more advanced since we travel in the universe. They have an effect that lasts for a week, which is why we also have a solar workshop on our ship.

Our solar-powered magnetism works exactly as the name suggests, it's powered by solar energy, just like you do. But the difference is that we can't capture pure solar energy like you can. That's why we have to produce solar energy ourselves, which is done through a solar workshop. But it's not a solar room. It's a workshop where we use a technique that absorbs the power of the sun. It is through this that we can charge our magnetism.

That's how it works on our ship, that we can navigate and propel ourselves forward using helium. And we know that you will also further develop our technique, but with what suits your planet best. Once you have learned this and adopted helium in the new sphere, you will introduce yet another new element. It will become a foundational concept within the magnetic field.

PROTONS – FUNDAMENTAL PRINCIPLES

Archangel Michael speaks:

To further increase the inner power, we also need a new composition, and in this case, it's about the atmospheric environment. This means we need to create an atmospheric environment so that we can combine our helium (and also water) with protons, with the power that has been formed through the same fusion. It's a composition that must happen after we have removed the electrons from the outside.

What we fuse the protons with is called a *hyperactive prepara-*

tion, which involves a mixture of protons, helium, and water. The helium is what transforms the proton's power, while the water only serves as a component. This is what the Kathremes were talking about – how they use helium, water, and other substances along with solar magnetism on their ship.

– How can we increase the proton's power? I ask Michael.

– We do this through the activity that helium consists of. Think about how it would negatively affect our world if we didn't learn to use helium even within the scientific nature. Because helium, according to us, is otherwise quite a cautious element – if handled that way. But through a composition, we can thus increase the proton's inner power thanks to a composition of helium.

When we then add helium and water, we dissolve the proton's inner power, which is reinforced by the helium's content. Then you create a base formula that you can also use for other things. The water only serves as a component, like a storage cabinet, we could say. Helium can thus increase the proton's power upon contact with water, which is what we want.

By using water and helium as additives, we can also remove electrons and neutrons from an atom, resulting in only protons remaining. The hydrogen gas gathers and becomes stronger, and the stronger it becomes, the more protons are formed. In other words, the strength increases as the number of protons increases.

– How can they increase in number?

– They increase because of the composition that occurs, because of the process that happens. Think of helium balloons. They can't float away until they have assumed a new composition of power. The same applies here. It will be used as a way to combine protons into more possibilities than we already do. Consider it a base formula on which you can then further develop everything.

– For example, into what?

– To act as a binder in an electrical atmosphere – to bind elements together. For when we bind elements together, the possibility that we gain *new acquisitions*, as we call it, also increases. A proton can move in all directions. However, what it cannot do is act as current or power. This is the task of the electron since it acts on the outside. It is thus the electron that controls the proton's progress.

So, the proton does not have the same strength as the electrons because they act inward while the electrons act outward. But if we remove the electrons, those on the outside, then the proton remains as the sole owner of an atom. When we do this, we introduce an entirely different movement, an entirely different power, that is, when the protons take over all the energy, and then we can transform them. It is actually possible to transform a proton into something else.

We can, for example, transform it into a current-carrying circuit. Almost like a circuit board with various associated components, but in this board, the proton remains as the sole carrier. And it is here, in this circuit, that we can increase the proton's inner power, which then emerges as a free-flowing force. A kind of transformation of a circuit board that transforms itself and then just keeps going and going. A kind of eternal cycle created solely by protons.

When you find this new transformation of the proton's power source, for example on a circuit board (circuit cycle), you will also find the proton's free-flowing force.

We can also use it in a catalyst and to further develop the power of the sun, solar magnetism, as iron is also attracted to helium and water.

We can also extend it to capture other protons that can be combined, which then contributes to it becoming stronger, and so on.

We know that helium, water, and protons are not new on Earth, and that's not our purpose either. The purpose is to further develop these three fundamental components, which your future students will do when they conduct various experiments with different compositions of protons. Because when the proton's power is completely unleashed and constitutes a completely unique combination of forces, you will also perform various combinations concerning protons, as you will then study them outside your school. It is then, when no one "controls" your projects anymore, that it will positively affect your results. It is also then that major changes in this field will occur.

That's why we have already placed many wise souls, though a bit arrogant, to work within this purpose because we need to go beyond the established boundaries of the school. For that reason, they may seem a bit arrogant at first, in their way of acting. The reason they have that attitude is so that they can develop new types of formulas and not always listen to what society thinks, for example, how or with what they should construct or combine protons. So, accept these souls, despite their seemingly arrogant attitude, because it is they who we need to be able to push forward so that we can move forward on this issue.

THE SECRET OF THE PROTON

Archangel Michael speaks:

Why is it so important that we develop the power of the proton? Parts of the proton's development have been kept secret, even though they never really were. Some research stations on Earth don't have the correct information regarding proton development. That's why we're bringing this up in our book—we want you to understand that protons are what you'll be further developing in the future.

– What do you mean by the proton's secret? I ask Michael.

– By the proton's secret, we mean the part that has been hidden from you. There are those who don't want you to further develop this concept because then you won't progress. What has been hidden is the part where we link protons with other underlying elements – it's possible to connect them with other projects as well.

They've also already developed this type of fusion themselves. This is what they plan to use for a future project without sharing it with others. That part of the proton's secret archive is included in their work, where they intend to open up a whole new project to then take over other research stations, which, of course, we cannot accept. That's why we're stepping in now.

It's a project they intend to use in space travel, where they harness the power of the proton. What they're doing is preventing other projects from using the same technology, thereby hindering others from making similar journeys in the future, especially after they've taken over other research stations.

– How can they do that?

– We mean it like this: if you master a specific type of travel technology, like a car, you can prevent others from developing the same ability. By doing so, you hinder others from making similar journeys in the future because the technology remains hidden. They'll also keep it hidden as long as possible.

– What technology are you talking about?

– The technology they're using allows them to travel farther into space. It's an external engine that lets them go even further into the universe. This, of course, cannot be allowed because everything we do must always be shared with everyone.

– Why are they doing this?

– They're doing it to monopolize space travel like no one else can. They believe no one will ever find out what they're doing, and they can then develop this proton technology on their own without sharing it. That's why our friends, the Kathremes, took a longer journey some time ago. They made sure to monitor what they were doing so they could pass the information on to you.

– Bermuda said they made a trip to the sun.

– Yes, exactly. It's through the sun's power that they've acquired this newfound proton technology, and it's also what they want to further develop and use to travel even further into the universe. And so far, everything is fine, but only as long as we share all work with others, which they're not doing.

– What can we do about it?

– What you can do is try to develop the same technology they're using. It's a brand-new proton technology they're on the verge of developing. That's why they made an open attempt at an external journey so no one would suspect anything.

But when we communicate, we never provide direct answers. We always convey the underlying thought they had with the journey itself. That's why our friends, the Kathremes, will continue to keep a watchful eye on them, so they know they can never rest easy with their secret.

This is why we'll be using more helium in the future. Helium is part of the foundation we'll use as a carrying medium when the proton's "full" power is released. Because if we release protons, we can also use the movement as a free-flowing force.

NEW ELEMENT

The Sun's Magnetic Union speaks:

To make everything work, we also need a new supply of material. It's a fundamental element that enhances particles, and that's what we want to discuss now. When it comes to magnetic particles, you will create a new formula of particles that can increase themselves with their own power, which you will then use in the transformer. Therefore, you also need a new power source for this. What we're talking about here is purely about one power.

- What kind of power is it? I ask the Sun's magnetic union.

- It's a summoning of power, as we call it, which should always be able to multiply—meaning it should always be able to increase its ability. So, for the particles to increase their ability, we first need to provide ourselves with a new element, in this case, a new fundamental material, which we talked about in part one. It's always important when we evolve to either obtain something new from outside or create something new ourselves.

When you reach this new fundamental element, you will soon notice how easy it is to control, which is exactly what we want. By "control", we mean that if it's easy to control, it will also be easy to create whatever you want to use it for. So, if you want it to multiply itself, that's what you create it to do. If you want it to soften rigid metal parts, that's what you create it for, and so on.

– Would you say it's versatile?

– Yes, exactly, but with a small twist. It can only be stretched to a certain degree, and you'll understand what we mean once you get there.

– How can we use the element? Wasn't its neutrinos that, by increasing their ability, would affect the speed in the transformer?

– Exactly right. What we mean is that neutrinos increase the power – they can even double it. But in this case, we also need something that can multiply the power that neutrinos create so that the power just keeps going and going.

– Do you mean that the element itself can double the neutrinos' power?

– Yes, but not exactly. We have to distinguish between them because they work in different ways. This element only comes in as an additional power source after the neutrinos have doubled their own power.

We can't go further than this, and we understand that it might be confusing when we can't clarify more. But trust us, when this new fundamental element comes in and you start doubling the powers from it, you will understand what we mean.

– Why are you bringing this up now, in this part?

– We're doing it because once you've gotten this far, you'll soon notice that the transformer isn't behaving as it should – not as we've explained. That's what we want to avoid. So, to make the magnetic concept work as it does for us, you also need to include and develop this element. It's only then that you'll understand what we want you to do in this matter.

So, find this new fundamental element and further develop it within the magnetic concept. Then you'll also achieve the extra power you need in the transformer.

THE ELECTROMAGNETIC COMBUSTION SYSTEM

Alfredo speaks:

When you've reached this fundamental element and learned about the Sun's energy levels, the power that the transformer will harness, and you better understand how we can link protons with helium and water, we need to create a combustion system in miniature form.

A combustion system is a way to burn energy, in this case, electromagnetic energy.

– Why do we burn power, specifically electromagnetic power? I ask Alfredo.

– We burn it to increase the power and to filter out any impurities, so that the weak particles, the ones that haven't successfully transformed their energy correctly, don't enter the transformer. We burn to increase but also to sort. A combustion system is, therefore, a way to alter an internal force, and that's what we use when we manage a current that needs to go in all at once. It makes it more resilient, we could say.

Because when we let the power pass through a system, in this case, a form of combustion, we can not only increase and sort the power, but we can also control it so that the particles give the transformer the correct power. Therefore, we can increase, sort, and control the power, and we do all of this by allowing the particles to pass through a combustion system. For this reason, we need something that can monitor the power, you could call it.

The particles will also be able to monitor their position. What we mean by that is that the particles will monitor the path they take inside the transformer because it's their path, and they never deviate from it. If they did, it would be like a car running off the road. Everything stops. Think about that. So, the particles can make sure they never go off the road because they are self-guided.

– How can they be self-guided?

– They are and can be because particles always follow a certain flow and never deviate from it because, as we mentioned earlier, they are drawn to it. For when particles work, they are always processed by an external force, in this case, the Sun. Therefore, they are drawn to the same link and never deviate from it. But if the power becomes too weak, they are eventually forced to deviate and end up off the road. This happens because they constantly need to divide to increase and continue their journey forward. That's why we need a combustion system that ensures the power is transformed correctly so that they never deviate.

– What differentiates today's combustion systems from the combustion systems of the future?

– That's a great question because it will make a big difference once you reach the concept of the future. In the future, you will be introducing a new type of electromagnetism, and it's also important that the combustion systems of that time can manage the components you will create then.

When you use today's combustion systems, you do so because you want to increase power and filter out impurities – you burn various components. But in the future, to burn electromagnetism, you first need to understand how we manage and transform power. Otherwise, it won't work. If you do it according to today's concept, you'll just get incorrect results because you won't be using oil or gasoline in the future. Therefore, you'll be sorting and burning different components.

– So today's combustion systems wouldn't recognize the way future systems work and the particles we have today?

– Exactly. However, you can use today's measurements, but with a little twist. When you measure today's components, you usually measure how much has been consumed within a single cubic unit. In the future, you'll measure the distance that the transformer has within itself. The power that the cables manage and transport will instead be measured in amperes.

As a result, your old system will collapse, but only to rebuild and create something new. Because you can no longer use a combustion system based on elements you no longer have, such as oil and gasoline. That era will completely end. So when you measure the future's combustion, think about how much the transport distance can accelerate into amperes, into power, how it can increase power while simultaneously transforming itself anew.

Does that sound confusing? Because that's what a combustion system will do in the future when you measure power. And if you do that, you can also see where the power is strongest and which parts need more power or don't need it, and so on

This is a concept we call *the electromagnetic combustion system*, which is essentially about increasing, sorting, and transporting electromagnetic power through a combustion system. But for the combustion system to work practically in your future feedback system, we need a unit that you can connect to the parts we need to burn.

Why do we do this? Well, to be able to burn the components and particles you'll be using in the future, they first need to be stored in another adjacent instrument. We do this to ensure that the particles that are transformed can be burned.

Imagine how much it would hinder us if we couldn't burn some components when they're transformed, and how chaotic it would become then. That's why we need to burn the particles that can be used, instead of, as you do today, where you burn to combine different components to create a new process.

So, in your future combustion system, you should only burn the particles that have failed to transform into the correct power, which happens through a type of transformer that can monitor everything to ensure it's done correctly. That's why we need a combustion system that can sort and control. Because if we don't have that, we can't see what's happening internally in the process where everything is transformed.

– So, it'll be a completely new type of combustion in the future?

– Yes, because when you replace the parts you have today, you'll also need to replace the methods you use today. That's why your old concept will be scrapped, and it's necessary to do that when you develop something new. So instead of burning, for example, gasoline, coal, oil, or other things, you'll burn particles that haven't been transformed correctly. This is done to separate the wheat from the chaff so that you only use particles that can drive things forward.

Once you've done that, the transformer will be exposed to a higher concentration of particles, and it's these particles that will then affect the speed. We can't burn at too low a speed because no conversion happens then. Conversion is when we shift to a higher energy frequency. That's why the particles increase their speed and contribute to a purer solar magnetic capability.

– How can it affect speed if the particles have been transformed correctly or not?

– They lose their power when they enter their early transformation, just before they've started their transformation. This is what will happen with all particles in the future. They lose some of their power just before they transform, only to then divide and increase their power again. This is what we mentioned earlier. In between, there's a combustion process that can separate the particles, those that haven't transformed correctly.

– So in this context, burning means removing?

– Yes, exactly. Or you could say that the combustion system of the future will sort and control particles, because that's what they'll do. Because when we burn particles, we increase speed by removing the particles that haven't transformed correctly.

Then we need to add one more process before all combustion can take place, and that's the flow. We need an even flow because the energy needs to be sufficient to power the transformer, which in turn drives the train or the car. So without a constant flow from the combustion engine, errors can occur. Therefore, you also need to build a new engine where all combustion happens, and the particles that the engine then sorts and controls.

This will be an engine powered by the Sun's energy. It will be the star of your project. Because without something that sorts and controls the flow of particles, it will be difficult to keep a car running indefinitely with its own universal power.

When you build the engine, make sure all the parts fit. Because in the future, we will no longer work with large packages to solve a significant movement capability. In the future, you'll work with micro-components where each part, for example in the transformer, can act independently. So initially, the combustion engine will be relatively large, but only as a starter package to then reduce in size. Everything will happen on the smallest scale in the future.

CODED MACHINE RECEIVER (CMR)

Kazandra speaks:

You also need to learn how to decode. It's not just the power of the sun that you'll be harnessing in the future; you also need something that can capture codes. In this case, we need a coded machine receiver. A coded machine receiver is what captures the codes that the machine transmits, and for the transformer to function properly, we need a coded machine receiver (CMR).

CMRs have been used for a long time in other contexts and applications, but in this case, it refers to a system that can re-code and decode when necessary. We can't just rely on the sun's magnetic capability to always function correctly, as solar flares can sometimes occur irregularly. Therefore, it's crucial for a magnetic concept to not only convert, distribute, and bifurcate energy but also be able to re-code when needed.

– How do we do that? I ask Kazandra.

– You do it by installing a decoding program in a sensor. In this case, we mean a sensor that detects the codes you've programmed into a software. But the transformer also needs something inside it that can read the software. Otherwise, it won't work.

– Wasn't that the responsibility of the integrated circuitry system?

– No, in this case, we need to establish it in a software program to be able to re-code, decode, or perform other actions. We need something in the transformer that reads the codes, the conversion of particles, and then sends all the information to a software program. It's a decoding system that's already available on the market but under different names. Here, however, we're only talking about those that detect the conversion, the composition of codes within the transformer. We call it *the transformer's navigator*, even though it deals with codes.

The Transformer's Navigator

Kazandra continues:

In your future transformer, you'll also need to implement a decoding system. This is done so that the part that transforms itself won't be disrupted. It will function as a safety mechanism. When particles are introduced at an increased speed, at the tremendous speed we anticipate will occur in the future, the risk of errors also increases. Therefore, you will install a new decoding program in the transformer, which sends out signals if something goes wrong in the process of converting particles.

– What happens if the particles aren't given the correct force? And how can a decoding system help us with that?

– What happens is that there's a sensor in the transformer that detects if an error occurs. The decoding then sends the same event to a connected software program that shows on the screen that something is going wrong.

In the decoding system, there will be different concepts you can use to solve such errors and decode or re-code so that the particles that went wrong end up in the right place again. The ability to do this is because there's already a setting in the sensor that tells it which conversion should be distributed. So if something goes awry or falls out of line, the sensor detects it and sends a signal to your software program.

But for the sensor to detect what's wrong, it also needed an unusual power. And since we're talking about a future electromagnetic capability, we can no longer use old sensors. They simply don't detect that type of change. Therefore, it's crucial that you create a new thread (sensor) in the transformer that has learned to detect when something goes wrong.

– What does this have to do with coding?

– It's called codes because it's codes you'll be inputting into your future software program. It involves different codes, conversion systems, and different forces of that conversion.

– Can't the code always consist of the same force?

– No, that's exactly it. The transformer on our ship consists of various determining factors. So one factor, one event, contributes a certain type of activity, and another contributes another. Sound confusing?

– Yes, a little.

Let's take an example:

We have three girls jumping rope. On either end of the rope stand two girls turning the rope while the third girl jumps in the middle. How fast the girls turn the rope determines how quickly the girl in the middle jumps. If they turn it too fast, it's because the inflow and outflow of force from the girls are too rapid, and then we need to re-code the force so that the speed slows down. If we don't do that, we'll have a faster pace than the girl jumping and the transformer can handle.

– So it's the inflow and outflow that determine whether the appropriate force occurs before it enters the transformer?"

– The force, just like the girls, is already there. It's the girl who jumps that we sometimes need to transform so she can keep jumping at the same pace as the particles being transformed. But sometimes, we also need to decode the force if it becomes too strong, and then re-coding occurs in the external field. But when we decode or transform the force that's in the middle (the girl jumping), then we decode and transform another force, and so on. Do you understand, my friend?

– Thank you, now I understand.

– Good.

It's important that the sensor can detect the force that's in the middle, where most of the transformation happens, and during inflow and outflow, so that we can preemptively decode and transform the force anew. This way, all the force within the transformer continuously receives a balanced dose of transformation. In this way, both the girl and the transformer can jump a little longer. The forces affected by an error are then removed, decoded, and the flow is allowed to regain its force, which we do via the transformer's navigator, the one in your computerized decoding program.

CHAPTER 15

CURRENT-CARRYING SOURCE

ELECTRO-SPHERICAL CABLES

Fredzo speaks:

How do we connect everything so that all parts receive the power emitted by the transformer, generated by solar energy, and then transmitted to other parts? We need power cables, and not just any cables. These are electro-spheric cables that carry energy and various components, similar to how a magnetic field works. Through these cables, you can unite the electromagnetism of the future.

Electro-spheric cables are just a name for the type of cables you will create in the future, designed based on the power and properties of a magnetic field. These cables will be constructed following the principles of how a magnetic field function. But to create these conductive cables, we first need to understand how a magnetic field is structured.

These cables will *carry* and circulate energy between the parts. To ensure the energy circulates properly, we need a magnetic concept that the cables are built upon. Therefore, we need to create cables that can withstand heat. Initially, they will become very hot, which is okay since this first version is only for training purposes, to help you understand and learn about their function. Once you advance further, you can apply a protective layer of iron, including using Mercury's method of containing heat, as we discussed earlier.

However, these cables, unlike a magnetic field, cannot repair themselves because they are not self-sustaining. Therefore, you need to learn how to repair a cable yourself if it breaks. This can happen if you pull them too hard.

It's essential to find a good balance in this pulling and transporting process that the cables will perform. When doing this, try to observe how a magnetic field repairs itself, as your angels have already written about. Once you have built a protective barrier around them, a thin one, they can begin to function as conductors between your parts, just like your cables do today.

– How can the current circulate, and where does the power that the electro-spheric cables transmit come from? I ask Fredzo.

– The current is added from the transformer; that's where the power comes from. The integrated coupling system then keeps everything in motion. It's also linked to a computer that uses a schedule to have the correct information (we will discuss the integrated coupling system later).

Alfredo explains:

Keep in mind: When we transport energy through cables, it affects not only the internal system but also the external one. When we use the sun as a magnetic component that provides current, we are drawing energy from the outside. Thus, all the energy we draw from the outside also affects the inside. When we then transport the sun's energy to the transformer, we must manage the energy correctly. Otherwise, it won't work.

So, how can we transport the energy safely so that our electro-spheric cables don't overheat? This is what you need to work on when integrating external energy into the transformer so that all parts can transport, manage, and transmit the energy effectively. We do this through a concept we call *adding a strong magnetic approach*, which we discussed earlier.

This will be an internal concept that you need to continue working on – how you can manage and transport the solar energy that will become very strong in the future. Not the sun itself, but the solar energy you will convert in the future will increase significantly.

Fredzo continues:

To be able to transmit power, we also need a current-carrying link that can generate an effective energy flow that can be adapted with your remote-control system. Without it, your motorized parts won't know what to do next. A current-carrying link arises from a configured remote system.

This is also where you can integrate a welded computer program that informs you about where and when energy is being connected, and where and when it is carrying energy. Therefore, it will be crucial that you create a current-carrying computerized program. This program will monitor and inform you when the link needs more power, when the current carrier has weakened, and at what moment this occurs.

– How can a computer program monitor and function as a current-carrying link? Is it like a battery, then?

– No, not like that. What we mean is a purer concept, so your battery will be considered outdated. What we mean is that you need to create something in the internal motorized magnetic system, something that controls and manages energy in the same way a car battery does. Therefore, it's essential that you connect the current-carrying link with a computer program. We call it a *computerized link program* that controls the current-carrying chain.

– How, and with what, can we connect them to a computer program?

– You do it through magnetism. It is magnetism that controls everything for you. It's also what tells you when something goes wrong.

– How do we create this computer program?

– Your future stations will take care of that. In just a few years, they will begin investigating the production of a link-carrying program that can function as current for magnetism. This is also what you will develop further.

Once you have developed these electro-spheric cables, you will also open various cable stations around the world, which is important to do. If something happens and only one country has access to these cables, you won't be able to repair your cables on-site. Therefore, it's crucial that everyone develops the same electro-spheric cables so that you can get help no matter which country you are in. The cables will be launched worldwide, but each country will develop its part, even though everyone will be part of the same written concept.

Feel free to call on us, Kathremes, through Ulrika, if you get stuck in this process. It may seem a bit tricky at first because it needs to be developed in many steps before you have the right strength to pass between the parts. If the strength is too high, other parts break. If too little, it doesn't get enough power. So, keep thinking about this, and as I said, feel free to consult us if you wish. We are happy to assist with this.

CURRENT BARRIER

Kazandra speaks:

For the current to pass through the transformer, we also need to learn how to bypass a current barrier. Otherwise, there is an increased risk that the current will weaken. Therefore, you also need to focus on *how we can transmit current between different barriers*. This is because it is often in a circuit where the current needs to pass that a barrier arises. And remember what we discussed earlier, that magnetism never moves perfectly straight because various barriers always arise.

The barriers that often occur in a distributed circuit are different segments. These could be particles that suddenly lose their power or other factors that cause the power to diminish. This is called a current barrier, and you can intervene to prevent a current barrier from forming. It is possible to do this, and when you do, your current will never weaken.

This can be done in two ways: Either you use a current distributor, or you learn to purify the magnetism, as we previously discussed. It will be a distributor that can intervene and alter the passage when it reaches a barrier stop (a distributor is a device used in technology, especially in internal combustion engines. It has several functions depending on the context).

– Can't we just use a regular power cable? I ask Kazandra.

– No, for this type of electromagnetism to work in the future, we need to add a new current-connected barrier breaker.

– How do we create such a thing?

– Your scientists will do that. But we'll give you a hint: Don't try to connect them initially, because if you do, you won't be able to break the power that is supposed to continue. First, use a magnetic ability that can interrupt an event, a barrier.

When you progress further in time, you can use a more detailed power cable, which your future scientists will build or create. In our case, we build our motorized current distributors so they can integrate with magnetic capability. Therefore, we also need to create new strong poles that can meet that power.

Archangel Metatron concludes:

In conclusion, we want to convey a future note about our electromagnetic spores. Spores, in our view, are the part that is current-carrying and also what needs to be improved in the future. Because today's spore, the heat-exchanging phenomenon, does not reach all the way and does not provide humans with the amount of energy supply needed. Think of it as a spore in nature, but in this case, we are thinking in terms of technology, but with a continuous stream of power that can bring more power within itself, and so on.

Therefore, at first, there will be a lot of calculations because not everything aligns when it comes to the electromagnetic sphere and the current-carrying spore that supplies heat. It is precisely your heating unit that is at a standstill today, so it needs to be reviewed. So, try to see things with fresh eyes and improve your heat-generating unit, the one that operates through electromagnetism, and you will eventually find the right solution.

POSITIVE AND NEGATIVE POLES

Alfredo speaks:

In order for the power to carry and transport such a strong electromagnetism, you also need new poles. Today's poles simply cannot handle the power of the future. Therefore, it's important that you first learn how to manage the energy, because the power that these new positive and negative poles will encounter is much stronger than the power you have today.

– What do the poles look like? I asked Alfredo.

– When you manufacture them, they are white, but they turn blue in the light. So, when they turn blue in the light, you'll know you've reached the correct pole. If not, continue to develop them until they turn blue in the light.

Furthermore, you will also need to learn how to create durable poles. It is crucial that the poles you develop also function during other transfers – they must be unchangeable because then they can be used in more places than just in a transformer. For if they could change in the middle of a process, there's a risk that they could develop their own power, which is possible, but we won't delve into that now. And since the power in the transformer is so strong, we need poles that can function as entry and exit points.

They simply must withstand the power that is managed where the current passes. After that, it's crucial to also think through *where* they should be placed. The power will be so strong that placement will be critical.

There is a country that already possesses this internal knowledge, and together with that country, you will achieve good results. This is also how you will work in the future, in different groups but still together. This will be a joint project where one country develops these poles, while another knows best where they should be placed, as they already have an enhanced element they can use to test where the poles work best. Therefore, our advice is to collaborate in everything you do.

When you find the best placement, the poles themselves will calculate where the current can pass. This is also where, in the same connection, you will need a temperature carrier. Both will cooperate, and the computer program will inform you when and where the power is strongest. So, close collaboration between the poles and the temperature carrier is important.

After that, you will further develop the poles so that they can also be utilized in industry. They will then function as *drivers*, as we call them, which means they set the pace. It is also at this point that you will begin to distribute new power from your newly established power stations as your new poles take on an increasingly important role. They will become that important.

Hint: Think of blue plasma.

CIRCIUIT BOARDS

Kazandra speaks:

We also need to create a new circuit board. It is important that the circuit boards of the future can withstand the enormous power that the car will have. Today's circuit boards cannot manage the power of the cars of the future. So, develop a new circuit board that can withstand high amperage and the solar power that the particles convert.

– What does the circuit board look like? I ask Kazandra.

– It will differ from today's boards because they need to be encapsulated. If they are not, they will blow out due to the car's enormous power. But with a power-bearing concept where the circuit board can be managed and encapsulated, they will work together. Therefore, you need to create a new protective layer but with completely different components than those used today. You can no longer use the same materials you currently have.

Think of the diamond concept, then develop from that. Something that is durable and can withstand heat without melting or blowing out, that's our advice. The diamond concept will also be further developed in other future motorized parts where the power exceeds the capacity.

Here, too, early collaboration is crucial because each country that develops a circuit board needs to have access to all the information. So, you will develop an early circuit board together.

By "early," we mean that one element needs to be developed before another. For example, you need to develop the contents of a car at an early stage. Otherwise, you won't understand how power can be distributed between the different parts and how the current can then be distributed through the cables we mentioned earlier. Therefore, the right circuit board is so important, and you should develop it as a collective effort that drives the car forward.

MAGNETIC REMOTE CONTROL

Kazandra speaks:

When you have created your circuit board, you'll move on to a magnetic remote control. A magnetic remote control is a kind of on-and-off switch. It's what we use on our ship to increase and decrease speed. You could say it functions like the steering wheel of your car.

That's why the remote control also needs to have a steering function so that we can turn with it. If it receives enough information about its actions, it could eventually steer an entire ship. But for it to do that, we first need to store the information the remote control needs to know. This is what our captain, Alfredo, specializes in – how we can apply and store information in the remote-control panel we have on our ship.

– How can we combine magnetism with a remote control? I ask Kazandra.

– The remote control gets its power from magnetism, including the movement scheme we'll go through in the next chapter. So, we first need to explain how magnetic ability works before we can understand and explain how a remote-control works. The remote control is actually a process controlled by a magnetic process, which in turn is governed by the time you input. You then link that with the time ongoing in the time exchanger when we start and stop. We can't create a remote control until we have the current-carrying parts we need in a transformer, which is powered by magnetic ability (we'll discuss time and the time exchanger in the next section).

Alfredo will now take over and tell you more about the remote control.

Alfredo explains:

To develop a remote control, we first need to understand what it should consist of. It's said that it should be able to start, steer, and turn. So, we need components that can do that, including an on-and-off switch. But to control an on-and-off function, we need steering components, and to reach those, you need to learn more about the basic aspect of magnetism – everything can attract if it just has the right charge. Because if we don't charge and set up a remote control with the steering qualities we want it to have, the magnetism won't know which direction to pull. So, we need magnetism that is cooperative and understands your commands.

For example: If you take apart a remote control, you'll see that it's also divided into different stations. Each station has its special property, but together they can act as a steering unit. Otherwise, the remote control won't work. So, take what you can from a remote control and then try to understand how you can apply the same technique to it, but with a magnetic approach.

How do we then create a magnetic remote control?

- We need a magnet.
- We need something that can start magnetism.
- We need something that connects the two so they can work together.

These are always the three basic requirements for creating a magnetic remote control.

When you're done with your magnetic remote control, you'll input the information you want it to execute. It's only now that you can input commands. So, if you want it to move forward, you input that. If you want it to turn off, you input that, and so on. And everything happens through computerized magnetic capability, so you can communicate with it. You give it instructions on how you want it to act. We also wouldn't be able to stop our ship if we hadn't already input a stop command.

This isn't a technology you have today; it's a technology that will be further developed by Swedish candidates because they're already advancing in that area. It's a new and fundamental concept that you need to start now, where you use computerized magnetism to give magnetism commands. This way, you can also skip the part where your magnetism stands still today. You'll understand what we mean when you get there.

You can also give it double commands, which prevents one-sided steering; otherwise, it could cause problems. And remember: When you give the remote control a double command, you also need to check why it only acts outwardly; it should also be able to act inwardly.

What else? Well, to improve its position, it also needs a well-controlled, self-steering navigation field. This is necessary; otherwise, your journey will be short. It's also where you can input quick commands in your navigation system, like "go there" or "go home," and so on. This is also a system that already exists and is well-developed for you. Here, you just need to integrate the same basic problem into your future magnetic capability. It will be a system that includes a keypad. So, by pressing a button, you can indicate where you want to go next.

– Will the steering wheel disappear, then? I ask Alfredo.

– Yes, but your autopilot will remain.

– So, in the future, I won't have to drive myself?

– Both options will remain for those who are interested. But most will be self-driving, and you can input where you want to go.

– So, some trips will be pre-programmed?

– Yes, far into the future. But initially, it will be voice-controlled, so you'll need a pre-programmed data system. You need someone or something that knows the destination.

– So, it's a self-driving steering system, but we program where we want to go?

– Yes, exactly. So no one else can control that part for you.

– How will we know how to program it?

– You'll learn that in a school. So don't worry.

Archangel Michael enters with a sharp warning:

Listen up! You should NOT use today's voice control; you must develop a new one. This is a technology they intend to further develop. You'll understand next year. Because it's a technology they've implemented, and it doesn't have clean intentions. Not the technology itself, but those who handle it. Therefore, you need to create a new technology that introduces an entirely new system, one that CANNOT control what you do or where you go. This is a warning issued by me, Archangel Michael. So create something new, and DO NOT use their old voice control system!

Alfredo continues:

To steer as you do with a steering wheel, it also needs to know which direction and how much it should turn. There's a big difference between turning right at a 35-degree angle or a 70-degree angle. This will also positively impact your development within the car's magnetic sphere. But that part will have to wait because it's in a more advanced line, but we'll gladly revisit it if you want.

A remote control also needs to be able to improve its functionality, and to do that, it needs to be created with a faulty link. Does that sound backward? Because it's not. The faulty link is supposed to function as a search engine if something goes wrong, which is important because all magnetism can always act independently. So, when you've input the commands, it should then be able to act on its own. We call this *an automatic system reset.*

An automatic system reset is the degree of reset that needs to occur when we check that everything is functioning as it should. That's why you also need to study a system reset together with a magnetic movement component. This is already included for you as a layer – how you reset a system when you need to review and check that everything is working correctly. Or if you need to troubleshoot something, which will then come up as a notification on your computer screen. You can then reset your already input concept, which then searches for an error on its own.

We know there's already an electronic search engine that finds faults in a technical process. But this will be a more advanced search engine because it will act when your cars run solely on magnetism, and that's when you'll bring this type of remote control to life.

You'll also need to add different concepts, along with magnetic movement capability, that today's search engines can't handle. Therefore, you'll also add a control system to check that the magnetic remote control is functioning and powered as it should be.

In the end, everything will end up in an automatic system reset, and it's essential that you think about that even at the beginning of your development. Consequently, it will be a single button that handles all resets, where it itself, when needed, goes in and resets everything.

This is a new type of remote control, not the one you already have. So, we're only talking about a future remote-control system that can be adapted to all parts. That's why it's important to distinguish the talent of today from what will come in the future.

We can't go further when it comes to automatic system reset, but we'll gladly revisit it if you can't find the right system for the right reset. Here, we just want to emphasize the connection between a system resetter and its magnetic capability, and how it can circulate in an internal concept. It won't be the same as you have today because you'll be operating in a completely different concept in the future, but the basic problem and execution remain the same.

Kazandra concludes:

In order for your future remote-control system (the concept) to integrate with magnetism, you first need to develop the magnetism of the future. The kind of improvement that happens between magnetism and electromagnetism can only be fully realized at that point.

After that, you'll learn how to control a remote device through a computerized unit, and then you'll combine the two. But before you can complete your project, you must first understand what a unit and an integrated coupling system are. The latter will be governed by time. Therefore, you also need to expand your knowledge about time and space, which we will cover in the next part.

PART 3

KATHREMES – TIME AND SPACE

CHAPTER 16

INTEGRATED COUPLING SYSTEM

LEVITATION PRINCIPLE

Archangel Metatron speaks:

Levitation is divided into two types:

1. The spiritual phenomenon – the ability to hover.
2. Magnetic levitation.

To achieve this, we first need to understand the principle of levitation. The levitation principle means that time can always be attracted and is primarily made up of various stages, which are also called different waves.

We have divided the principle of levitation into two parts:

1. The first step in the levitation principle is the elemental part, which involves a simplified version of the levitation wave – like the ability to hover.

2. The second step is part of the multilayered line and involves a wealth of knowledge. Everything within levitation always consists of different stages. These stages can be set in time cycles, or time can be divided, and so on.

The principle of levitation is therefore a part of the universe's nature of changeability, because if things could not change, everything happening between time and space would be completely still. This is where it differs from researchers who have not yet accepted that everything operates in time and that time can also be divided, even within the same space.

So how do we develop levitation, for example, for cars or trains? This is where our integrated coupling system, aggregate, and transformer come into play, which we have already discussed.

We have divided it into three parts:

1. A transformer powered by solar magnetism.

2. An integrated coupling system controlled and created by the rhythm of time.

3. Time and motion aggregate.

INTEGRATED COUPLING SYSTEM

Archangel Michael explains:

An integrated coupling system is a movement schema in time, and to explain it, we have divided it into three parts:

- Integration of a space in time and space
- Integration of a computer-controlled program
- Integration of a remote control

  = Integrated coupling system

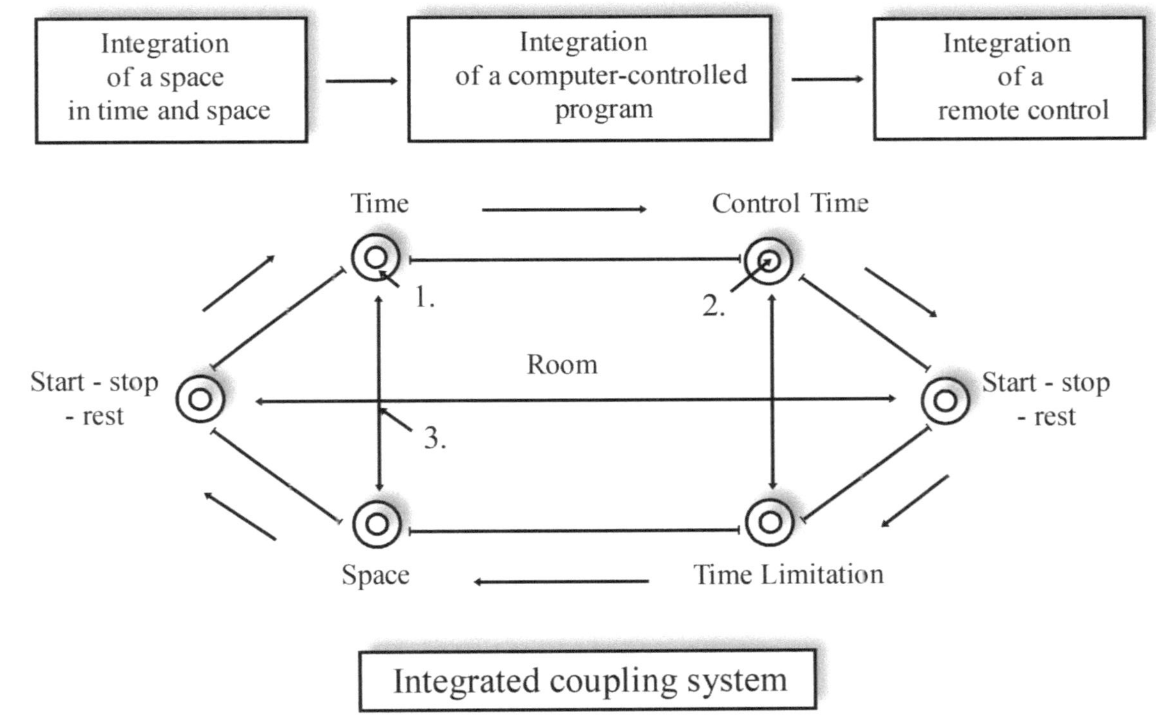
Integration
of a space
in time and space
Integration
of a computer-controlled
program
Integration
of a
remote control
Time
Control Time
1.
2.
Start - stop
- rest
Room
Start - stop
- rest
3.
Space
Time Limitation
Integrated coupling system

(Together with Archangel Michael, we created a simplified illustration of what an integrated coupling system looks like. Time and space, time constraints, and other factors are just examples. The positions in our illustration are not exact. You will need to adapt them to what works best for you.)

See illustration on the previous page.

1. Intermediate Layers and Various Coupling Stations – This is where all the charging occurs.

2. Gate – Here you can calculate the smallest details of what happens between time and space.

3. Navigation Field – The space you have to navigate within.

Space = The space will expand depending on resistance. Get to know the surroundings, and you will also understand each resistance present in every space, which will help you determine the capacity to use.

To achieve an integrated coupling system, you first need to learn to calculate time. Therefore, you need a controlling, time-determining aggregate that you connect with a magnetic capability. This means that if you need a coupling system for a time machine, you create a schema for it. If it's for a train, you create a schema for that, and so on. Thus, we need to establish a schema for the specific space where time flows. This is the only way to activate an integrated coupling system with the help of magnetism.

AGGREGATE

Archangel Metatron speaks:

An aggregate can be a motion aggregate, such as a motion schema as mentioned earlier. There are also time aggregates, which we will discuss later. Aggregates can be found at different stations.

An aggregate can also be used to study the difference between *effect and uneffect*, as we call it, in what happens between time and space. An effect is what occurs – a change. An uneffect is when it feels like time stands still.

Different events need to be connected and integrated into the aggregate you wish to use.

An aggregate is a crucial component, and without it, you cannot progress, no matter which space you are in, or how slowly or quickly time moves. Consider how essential gravity and magnetic fields are to the universe. Without these protective aggregates, other parts of the universe could disrupt your journey and the time you operate in on Earth. Just to give an example.

But don't worry. Future researchers will develop different types of aggregates for that purpose. We have already placed them at various stations and schools, and they will meet in the future to share their knowledge. There will be grand gatherings where you can learn from each other. It will be a mix of both teachers and students, which we always consider the best combination for development.

You will also reach a universal aggregate in the future, but not until you have learned to use the universe's power as a sustaining source instead of the electricity you use today. At that point, you will need a different aggregate to set everything in motion.

It is also then that you will learn how to control time, which we will discuss later. But to crack that code, you first need to understand how everything occurs in time and space, how it has an effect or an uneffect. Once you understand this and have learned to time-limit a time aggregate, you will reach a better solution for today's forward movement.

So why is it important to understand and further develop our knowledge about time? Time and space are different concepts you will use in the future, including a time aggregate. For if you ever want to convert power, such as in a transformer, you need a time aggregate that can also do that. Otherwise, you won't achieve the right flow, so to speak, and the transformer won't function in all spaces of time.

The goal is to use time and space to create a motion schema for an integrated coupling system. Then, magnetism acts as a motor for our motion schema, and the ultimate goal is to create a levitation car.

This is what we will cover in this section. So before you fully understand what an integrated coupling system and aggregate are, you first need to grasp what time and space are, as that is where you will develop the levitation principle.

The universe does not create a magnetic effect
until it has created the parts that can understand
and manage a magnetic effect.

CHAPTER 17

TIME AND SPACE

TIME AND SPACE

The angels speak:

Everything in the universe has its time, and wherever we are, we are in a space, and for a space to exist, we need time. We cannot be in a space without time. If we were, it would simply be impossible to exist in that space. For a change to occur, time must be in motion. It is impossible to achieve a change if time is standing still.

Imagine being in a store, and suddenly, out of nowhere, both you and time come to a complete standstill. You become frozen and cannot move forward. That's how critical time is; without it, we would be immobilized. For time to travel and change, we need a space where time can pass. It is only in this way that time can begin to move again, through the existence of different epochs and spaces that we can traverse.

How can we understand how everything in the universe and our own world integrates if we don't understand the time and space in which everything operates? This is where we ultimately end up when we seek to improve our technology and fuel systems.

It is, therefore, the knowledge of time and space that we must attain first before we can achieve new and improved possibilities.

– How can the soul provide us with knowledge of how a coupling system works in time and space? I ask the angels.

– We need to understand ourselves first before we can understand the universe, and we need to understand the universe before we understand how an integrated coupling system works.

– Aren't we already doing that?

– You are, but not entirely, as much information has been omitted. Especially the part that constitutes the small details, the ones we discussed earlier.

So, to travel between time and space, we first need to understand where we come from. This is something some researchers try to skip, as they have neither understood nor had the patience to let time take its course. It is crucial that we do not jump ahead but allow all development to take its time, at the pace of human awareness. Since all development always occurs within a certain framework and time, it is important to follow the line we are on.

Therefore, we wish for you researchers to release that aspect now – traveling back and forth in time. We currently have other important tasks waiting for you at your desks that require your full concentration.

Instead, our more media-sensitive souls will flourish in this subject—those who have consciously chosen to develop their full potential in this area because it is needed. As future and past journeys can currently only be performed within one's inner self, it is only the highly media-sensitive souls who possess this knowledge today. This knowledge allows certain souls to break the barrier that separates time.

THE SOUL BREAKS THE BARRIER

The angels explain:

To fully understand time and space, we first need to understand ourselves—the soul. However, in this section, we will focus only on the motion schema that exists between time and space. Thus, we will only highlight time, space, and the soul in this context. The soul's journey is a deeper exploration of time and space, which we might further develop in the future.

As always when discussing the universe, we first talk about the soul because the soul is a part of the universe. To break the timeline that separates the present, past, and future, humanity needs to climb mountains you have yet to learn to scale. This requires enormous knowledge and a soul that is well-versed in these concepts.

There is a linear line that separates time. To break it and travel back and forth in time, we must first learn about astral travel. This is where you find all the foundational knowledge needed for further research into time and space. Astral travel involves an experience where a person feels as though they are floating outside their body.

Let's use an example of how we angels break the line between all the spaces of time:

Some people will always question how we can be present with you simultaneously. The answer is simple: We break the line that exists between time and space. When we move quickly, it feels like time stands still, and when we move slowly, it feels like time speeds up.

– Sorry to interrupt, dear ones, but for me, it seems to be the opposite. When I am active, it feels like time flies, but when I sit still, it feels like time drags. How can this difference occur? I ask the angels.

– We understand your perspective, and many people feel that way.

Imagine you are walking down the street, and suddenly everything freezes. People turn into statues. Everything except time stands completely still. But there is one man who can still move freely. The man is then in a better flow than the others because when everyone else is frozen, it's easier for him to move unhindered. But when everyone is moving, some of the free space he had before disappears. Thus, it becomes harder for him to move.

For the man, time seems to pass quickly, but for everyone else, it feels like time is moving slowly, even though it is not. Time is still moving forward; otherwise, the man would not be able to move. This is somewhat how it works for us as well.

We understand if this is challenging to grasp, but if you think about it, why wouldn't it work this way? There is a reason why great scientists like Einstein and Stephen Hawking discovered this science because, according to us, it is useful. Time and space can indeed be altered, and this is what we use when we operate with you simultaneously.

When you move, we also need to do so to keep up with you. It feels like time moves quickly, but in fact, it does not. When you are not moving, time seems to pass a little faster for us, even though it feels like time is standing still for you. We can then, as mentioned, work more freely.

– Is that why I sometimes feel an inner stress even though I am sitting completely still? I was once told by a medium that when you shift positions in my channel, it feels like my thoughts race, and I can't switch off.

– Yes, that's right. We are constantly changing and trying to adapt to your inner state. But sometimes, we also need to move, both to change our position and to determine who should be by your side in your channel, which can naturally affect you and your energy level. So, to reach you, we need to be here and now, where time is in motion, even if it feels like time is standing still. This is why it sometimes feels like the pace quickens and time speeds up when we are in motion, even though you are completely still. Therefore, time, space, and the present moment are important when we communicate. Do you understand, Ulrika?

– Thank you, I think I do.

– We understand it can be difficult, but that's how it works in the spirit world. We will continue with the present time.

PRESENT TIME

Archangel Michael speaks:

Why is it that we can travel between time and space while humans cannot? It's because there is no "now" in the universe. The present moment is just a concept human use to function on Earth. If humans lived in the universe's timeframe, without the present, you wouldn't be able to be present. The universe isn't present; it simply is.

So if humans lived without the concept of now, you would also just exist, which works for the universe but not for humans. If you weren't present in time, you wouldn't be able to live *flexibly*, as we call it, and there would be even more chaos in your inner world than there already is. You also wouldn't be able to act as a seeker because if we just exist, we wouldn't care about what happens in the future. This is why humans exist in the present time.

However, for the universe, the present isn't important because the process happening there is always moving forward. The universe never stands completely still, as humans sometimes do. Not time as you experience it, but sometimes your physical bodies stands completely still. So if there's always forward movement, you can never live in the present. That's why there's no present in the universe. That's why we can bend the concept of time here and be more flexible with it, because we simply exist.

– What happens then to time and space when we communicate with each other? I mean, how can we bend time and space so that we can interact with one another? Aren't you in a completely different dimension than I am when we speak? And is the information up to date? I mean, does it happen in the present, future, or past? I ask Michael.

– When we communicate, everything happens in the present. But you're right, sometimes we are in a different space; otherwise, we couldn't reach each other. We mean that if you're sitting in your kitchen writing down everything I'm conveying, both your physical and spiritual bodies are in the present. But sometimes, as you might have noticed when communicating with the Kathremes, it feels like you're not sitting in your kitchen but more to the side. That's because you're in different dimensions. And for you to communicate with each other, you needed to meet halfway, which means you elevated yourself a dimension, and they lowered themselves. That's how you made contact. Do you understand, Ulrika?

– Yes, thank you. But are we in the same time and space even though we are in different dimensions?

– Yes, you are. You're in *a central position*, as we call it, which is crucial; otherwise, you wouldn't be able to reach each other.

Think back to a time when people couldn't communicate over long distances. It was impossible to talk to someone who lived far away. But as technology developed, it eventually became possible. It wasn't possible as long as the central part wasn't there because we always need a starting point.

It's the connection we have that made this possible. And as you've opened yourself up to process your inner self, your central part, you've done it over long distances as well. The more you process your inner self and understand yourself while simultaneously developing your spiritual technology, the longer and better your connection becomes, for example, with the Galactic's.

– What does this have to do with time and space?

– It has everything to do with time and space because without time and space, there's no forward movement, and then we don't evolve. If time stands still, so does space. Do you understand, Ulrika?

– No, not completely.

– Let's try again.

Imagine you're standing in a room. Time is completely still; you're not moving forward. But if time, your development, happens in the same room, then we introduce a forward movement. Because without movement in a room, right where you are in the present, we can't move forward if time doesn't move. It doesn't matter where or how you do something when you're in a room. If time doesn't work, neither do you. That's how all development works.

You might be in a room in the present, but if you don't develop yourself, it feels like time is standing still because nothing new is happening. But if we learn and dare to try new things, we also move forward. So in both positions, we're in a room, but if nothing is happening in your life, even though time is passing, it feels like time is standing still.

That's why it's important to at least try to introduce new time, in a new room, in the present, because if you do, then time moves forward. And as you develop, so does time in the room you're in. In this way, you can actually influence your own outcome – whether you develop or not – by introducing new learnings. Because if you increase your inner self, you increase your hunger for even more knowledge – there's a continuation.

So without your fantastic development, it would be difficult for you to be in the same position as our friends the Kathremes, because it first required an understanding of your own inner self and where you were within yourself.

Those who don't understand this are often the ones who hesitate. That's why they can't take in the information we give them, because they haven't completed their inner journey yet. But by learning that it's possible to move through different dimensions to communicate with us in the spirit world, you also understand the time and space you're in, internally, when we, for example, write together.

I, Michael, am now on your right side, but at the same time, my soul can explode into many different dimensions. And because I've reached that knowledge, I can also understand and travel through all the rooms of time. It's a bit difficult to explain because it's more complicated than that, and we don't want to confuse more than necessary. But we know you understand and that you yourself have experienced the time frame that exists here, where we are and work together.

In this way, we can bend time and space more than scientists understand today, because there is no now in the universe. So when we understand the present time and where we come from, we can move on to understanding what time and space are. Because we cannot understand what time and space are until we understand ourselves, and when we understand ourselves, we also understand the universe.

GAP BETWEEN TIME AND SPACE

The angels speak:

What do we mean by a gap between time and space? Let's say, as mentioned earlier, that the spaces we need to pass through for time to move forward suddenly disappeared—at least half of them. We would need to jump very high and very far to reach the next space, to get to the point in time where we want to be. But when we skip over a space, we're also skipping over time, because if we're supposed to follow all the cycles of time and the changes that need to occur without causing too many negative consequences, we can't cheat. We can't skip over any spaces, but that's what some people do when they develop their technical or motor skills.

So before you continue developing your technology or anything else, you need to stay in the same space for a while longer, during that time. This is both to understand and to think about whether what you're doing now might have a negative impact on the future, both for the Earth and for humanity. But some people don't do that today – they develop without stopping, without thinking, and with each passing decade, things just go faster and faster. It's like living in an eccentric bubble where only you and your technological development exist, and there's neither space nor time left for anything else.

That's why you need to stop the train for a moment, go back to where you once were, and stop performing at the speed you're going now. Because the faster things move for humans, the faster they also move for Mother Earth since the two are synchronized. If instead, we learn to integrate our teachings and our energies with each other, we can save the Earth rather than destroy it, simply by understanding time and space.

– Why do humans skip over some spaces? I ask the angels.

– One explanation is that humans believe there's a gap between time and space, but there isn't. And if we're ever going to change the Earth, we first need to understand that there isn't a gap between time and space. It's humans who think there's a gap, and that's why you skip over parts that we don't.

Think of it this way: If nothing else in the universe creates a gap, why would time and space do so?

It's through knowledge of the universe that we can also improve the technical abilities we need to reach as far as the past, present, and future, which humanity has already done to some extent. But don't forget that age also affects how far we can see. An eye can only reach as far as it was created to, because there's a limiting effect in your eye that influences how far you can see into the distance. So how far you can see also affects how far into the future and present you can see, or how far you could see when you were a small child.

Aging weakens not just our external motor abilities but also the movement that occurs inside the eye. The slower the motor skills, the weaker your vision, and so on. The same goes for our technical abilities.

Why are we talking about the eye in relation to time and space? Back in the day, humans believed the Earth was flat because the mechanisms we had at the time couldn't give us far-reaching information. Today, we're a bit further along in time, and we've reached the point where some believe there's a gap between time and space. We can't see it with the naked eye, but since we've begun an inner and outer understanding of time and space, we're quite sure that there's a gap between time and space … or is there? Or could it be that some scientists' sources aren't entirely reliable on this subject?

We mean: The only ones who can see into the future or see what lies between time and space are those of us who exist in this time dimension. And because we exist between time and space, we know with absolute certainty that there's no gap between time and space. If there were, even we would need to jump over these spaces, which we cannot do.

For example, humans take measurements with the instruments and eyes that you have today. So our question is: Are they 100% equipped to gather information about what's true for the future? Or is your technical equipment better suited to gather today's information? We would probably say the latter.

Moreover, when you stop skipping over the spaces you're in, you will also understand the journey you need to make in the future to learn how to travel between time and space. Then you'll see that there isn't a gap between time and space.

Time and space are interconnected, so without time,
space wouldn't exist.
Therefore, there's no gap between them.

TIME CYCLES

The angels speak:

However, there are different time cycles that follow us along the way, but these are teachings in time, which is where you find yourselves now, nothing else.

– What do you mean by time cycles? Is it the same as a higher dimension? I ask the angels.

– We often use different time cycles when we talk about concepts related to time and space. But we can also say that we move forward into higher dimensions in all spaces of time as we increase our energy frequency.

– How can we live in different time cycles?

– You can and do, depending on where you are in your learning process.

– If we are in different time cycles, how can we still integrate with each other?

– It's not a physical time cycle. It's a time cycle for learning, where those who develop and learn from everything that happens leap forward to new opportunities and new stages of development. Meanwhile, those who remain in the old concept don't develop in the same way.

– But for some, that might be the case, right?

– Yes, and we understand the subtext of your question. That's true, and some need to be there to try and experience what isn't so good, so they can then grow and move on. So, you are only in different time cycles in your learning, not entirely physically. But as we adapt to a new era and begin to encounter a new world, not everyone can meet that time because we first need to be there in our teachings. And if you haven't reached that point yet, your soul will remain in that time cycle until you awaken and learn.

– How can time and space affect us like that?

– It can because the soul isn't affected by time in the same way as your physical body. And as you now know, there is no now in the universe because the universe simply is

– So, if we were also to just be, would that affect us positively or negatively?

– Both. Your physical bodies need to be in the present, otherwise, it wouldn't work because you'd become too confused. But the soul knows that it doesn't need an exact time or a now to learn and exist because it just is, just like the universe is, and the soul is part of the universe. This is what we discussed earlier.

– So despite our different time cycles, can we still eventually meet on the new Earth?

– Many of you, like you, Ulrika, have already started working according to the concepts of the new time because your souls are in that time cycle. So many will wake up and follow you, but not all. Some will continue living in the old footsteps for a while longer, and it's important that they get to do so until they too reach a new and higher time cycle.

– So everything happening now is for our development?

– Exactly, even though we understand that some might feel sad to hear that.

Think of it this way: How would we learn about justice if we never encountered injustice? How would we learn from the light if we never faced the darkness? How can we become strong and courageous if we never encounter discomfort and fear? How can we grow in body and soul if we never expose ourselves to what helps us grow?

How can we live in a wonderful family relationship and understand how it works if we've never experienced a broken family relationship? How can we understand the importance of a strong relationship if we've never been separated from one? How can we learn and understand that death doesn't exist, that no one can die, if we don't understand the purpose of the soul's journey? That it's your physical body that leaves you, but your soul lives on, and it's the one that grows and learns. Your physical body is just a tool that you use so your soul can grow.

That's why it's the soul and time and space that we want you to begin with now, a deeper version. Once you understand the soul and that there is no gap between time and space, you'll move on to the next phase in the same time cycle.

You will also further develop your work in the future, and only then will you change your perspective on life. That's when we'll enter a new and changed time cycle. Because when we change our way of living, we need a new time aspect, a new time cycle, that falls in the same linear line and synchronizes with the soul's development.

– So it's the development of the soul that determines how and at which timeline we end up in the future?

– Yes, exactly. That's why you can't leave the time cycle you're in now until you're completely finished with it. After that, we switch timelines, and that's when we throw away the old and instead create something new. That's when you'll increase your knowledge of time and space, and when you do, you can also learn to control time, which we will go through in the next chapter. But before we get there, Archangel Michael would like to speak first.

CODED IMAGES

Archangel Michael speaks:

It is also here, across different cycles of time, that "they" alter their material. Therefore, we will now conclude this chapter with something entirely different, but of great importance to bring up now to explain both what has happened and what continues to happen today.

There are those who have advanced much further in their technical development than you have when it comes to time and space. It's a technology they use to deceive. How do they do that? Well, to look back in time, we need a device that can reverse. For that, we also need a forum with different sections, where each one represents its own time. In this context, time means the past – what has already been.

But to break time, which we sometimes can do when looking back at the past, we can follow the pattern that was set during that particular event. That's how we can look back in time. We can't physically move our bodies back in time, but we can use the information that existed then and bring it to the present. That's how they do it.

– Why, and how do they do that? I ask Michael.

– As I said, there are those who have already developed a technique so advanced that they can see what has already been, both here and on other planets. Therefore, they can also use past information in today's context to make it sound credible to your ears, to deceive.

– What kind of technology do they use? I ask.

– They use a technique that has actually been around for quite a while now. What they do is capture systematic codes that they then distort, which happens through a systematic program they've created.

I'm going to give you images in your mind now. You see lines, kind of like a barcode that the cashier scans, where each line represents a code. That image, the code, they can move forward or backward in time. But only through a systematic program, because that's where they can break time. Breaking time in this case means either removing a part of the image that exists in that specific time, in that specific event, and instead inserting another image from a different time. That's how they operate.

– What do they gain from this?

– So that you will believe that what is happening is actually happening, even though it's not. That's how they scare people with images and codes that never happened or that happened in the past.

– How can we know this?

– You can't, because it's a technology that's not yet in your hands. But when it comes, you'll understand. The technology we're talking about is a more advanced one that can automatically find where each image has been misplaced. That's why it's so important that you trust your inner self, instead of what you hear and see around you.

– Why images in particular?

– They mostly use images because their words don't give them the same results. They can't feel, not like you can. Therefore, they don't understand what others feel. They don't understand how time or an event feels or functions. And isn't it sometimes like that for you that it feels hard, when time feels like it's moving too fast or too slow in the situation you're in? They can't feel that. So, they sometimes create without understanding the outcome of their own creation. The only thing they can do is create, but they can never feel the outcome of an event in that particular time. You'll understand what we mean. It's in this area where their weakness lies, and it's also this weakness that we use when we break their codes.

As they can't feel an event
that happens across all times and spaces,
this is what will ultimately
lead them astray.

As a help, you will develop a completely new systematic decoding program. It will be a program that you can use to break down the image and divide it into different time periods, allowing you to determine if the image is correctly placed in its time. You'll break down different points of the image to see where it comes from.

– How? I don't quite understand.

– You will work according to a point list, where each point in the image represents its own time of that part, of a micro part.

Think of it like this: | . | . | .

Draw points and between each point, between each event, there is a time – the lines. So for each event, time is broken down to see if it fits in that specific time, that is, if the image also follows into the next time. That's how your point list will look for a while before it's further developed.

Let's take a drawing pad as an example:

On the first page, you draw a man. Then, you change the man's movement on each page. In a decoding program, this is called a pattern. In this way, a change occurs in that particular event. When you're done with the whole pad and flip through it quickly, it looks like the man starts walking, and this is exactly what we mean when we talk about images from the past.

Because that's what they do when they use their encrypted program. They can take out a piece, tear out a page in the middle of the pad, in the middle of the movement, in the middle of time, and instead insert a new page, which you can't see. But when you've created this systematic decoding program, you'll see that the points that exist neither match nor follow the timeline they're supposed to because they've been distorted in time.

So, follow the flow of lines, that's our advice. You'll then reach the next event in time and can continue with the next in the same linear format.

We'll take yet another example, a mirror room at a circus. It's really important that you understand this:

Imagine you walk into a mirror room and stand in the center. Numerous mirrors surround you completely, and each mirror changes how you look because each mirror represents its own time, its own event. But together, they form one and the same event – your reflection. Through the program they use, they can remove a mirror image, the one between the points in time, in an event, and instead insert another mirror image so that you'll believe that the event is happening here and now. But it doesn't necessarily have to be. It could just as well be an old event that happened, but that they're showing today to make you think it's happening here and now.

That's how they deceive, and they know they need images in your reality for you to believe in what's happening, even though it isn't. You're standing there, you have the image right in front of you, so you believe what's happening. Then you react to what you see. And do you remember at the beginning when we mentioned: "If you can't see it, then you don't believe it. But if you see it, then you believe it." The same applies here.

– What do they use to change the text?

– They use source codes. It's a coding system that allows them to go in and change dates, times, and years, but most importantly, the text. That's the part we need to look into now to stop those who have this kind of alteration. Therefore, you need new tools, especially those who work in media. It's going to be those who will review some of the falsehoods, but only those who work in the service of good. And I, Archangel Michael, will grant them all.

– How can they get in touch with you and get help unraveling this?

– It will happen naturally when the time is right and by the souls who already have some information. These are the ones who will receive new keys. We first need to unlock the tools they need before they can legally go in and make changes. They will then unravel a lot of encrypted text, and their own channels of information will reveal themselves. It will be a time of great confusion because you've always believed in their words, but then you won't be able to ignore or deny their way of working any longer.

Furthermore, we can say that it's a long-term program that will take time to develop, but don't give up because it will be good. Once you've developed this program, it will yield good results within a week. Then, more false images will also be revealed. The program will start in one elongated country, but where the power lies in another. There will be some puzzles to solve initially, but you will understand as you go along. You will also continue to develop this searching program.

So, start reviewing images for those of you who feel called. We also have many artists on hand, so take their help as well. They will also be the ones to decode these images. And call upon me, Archangel Michael, if your words or the feeling for your work becomes stagnant. I will then show you on the screen how this is meant to be.

This is what many have as their life purpose in this life, to first make mistakes and then change and make things right, which is a fantastic path to take according to us. Therefore, many will now turn towards the light and help us through the time that is coming in the future, where they will explain how they have worked. When this happens, we want you to greet these souls with open arms and welcome them back into our world. For many have themselves been misled and should not be blamed for what they have done.

CHAPTER 18

CONTROLLING TIME

ATTRACTING TIME

Archangel Metatron speaks:

How do we control time? We do this by first learning how to attract time. How do we attract time? By introducing a faster flow of time within a space. How do we introduce a faster flow of time within a space? This always happens through different time cycles. It is within these cycles that time can change its nature. If time can change its nature, we can expand it, and if we can expand it, we can also attract it into a space.

– How, and in which part do we attract time? I ask Metatron.

– Only we can do this because it requires the ability to separate time in each space. This is how we can illuminate information that comes from both the future and the past. By dividing time, we can attract it, and by attracting it, we can also pass through it to move forward and backward in time. This is where the principle of levitation comes in.

We know it may seem complicated, but you will understand because the principle of levitation, that time can be attracted, will also be used on Earth. When you do this, your understanding of how light can be divided into multiple movements will also increase, as we previously discussed. This is a technique that you too will use in the future across all research fields. Therefore, everything related to the principle of levitation will gain a higher understanding, both for how time can operate in all spaces and how it can be applied even in the most challenging areas.

CONTROLLING TIME

Archangel Metatron speaks:

Once you've learned how to attract time, you can also learn how to control it. Controlling time is another topic we want to highlight now because it exemplifies the very essence of this concept – it seems as if time stands still, even though it doesn't.

Time itself isn't moving as many believe. It's everything around us that moves in response to time. Does that sound backward? To explain what we mean, we've broken it down into two aspects:

1. Consider surrounding factors.
2. Resistance.

A long time ago, people believed that the sun revolved around the Earth, but today we know that the Earth orbits the sun. Naturally, the sun also moves; otherwise, there would be no resistance in its motion, which would affect time. But for time to pass, we need something around us that is constantly in motion. It's not enough to simply understand a situation. We also need to learn that the sun moves; otherwise, it would result in a one-way motion, a one-way understanding, which would impact how time progresses overall.

Let's take an example:

Imagine you're swimming. A swimmer in the lead swims faster than everyone else. Therefore, you're affected by different surrounding factors than your competitors are. But even though you're far ahead of the others, you're still in the same space (the pool), which impacts your swimming technique because everyone's times or finishing times are affected by their surroundings. So, the more swimmers swimming side by side, the more they influence each other. A swimmer in the lead has greater freedom in their movement and can use time and the movement of the water in a way the others cannot. It is only those who are in motion who can influence how time moves, depending on *their* surrounding factors.

This is how we teach our students in the spirit world – how to use and understand time and space.

In the same way, we can learn to control time, as it positively impacts your work if you can remain in the same space a little longer. So, to learn how to control time, we also need to consider other surrounding factors, other situations, that are also in the same space, and how they affect your time. But there is yet another aspect to consider and that is resistance.

For resistance to occur in the space as time passes, we need something that pushes back. In this case, the space itself provides that element.

For example: If you're in a cramped space with others, it might not just feel like they're encroaching on *your* space; it also feels like there's no forward movement because you feel cramped. Therefore, it seems like you're standing still even though time is passing, but in reality, you're moving forward.

The space wants to expand because
it senses resistance,
but there's not enough time because it feels like time
is standing still.

Another example: You're standing still, alone in the middle, and suddenly the space starts to move. You now know that time is flexible; otherwise, time couldn't move. You start to feel trapped in the space, so you create your *own* internal resistance. If you didn't feel resistance when time and space are in motion, you'd move forward more easily. It's your own influence that determines whether you experience resistance or not. Instead, you can remove the resistance and free yourself. Only then will you move forward.

It's similar to your learning. You need to fully learn your lessons before you can move on to the next. If not, you experience resistance. You don't move forward in life, even though time is passing. And this is precisely the development we want more people to embrace now. Time is short in the spaces we occupy, and the resistance we experience is only caused by not having the information we need to move forward. But with more knowledge about resistance, you'll advance in terms of time and space. Then, you can also learn to control and break time. Once you've done that, you can overcome your resistance.

MIRAGE TECHNIQUE

Angel Fabrizio speaks:

How, then, do we overcome resistance and control time? We do this through the study of mirage technique. This is where you'll learn to leap over time and add a magnetic movement effect. But to create a mirage, you first need to understand how to merge different time concepts. There are four different time concepts we use, namely: dimension, illusion, selection, and prevention.

- *Dimension* refers to the various levels that exist in the spirit world and throughout the universe. These are the steps we must take to achieve a complete result. It involves different levels of power, different levels of information flow, and so on.

- *Illusion* is another word for mirage. But for us, illusion is necessary to understand different time concepts, which, in our view, cannot be achieved through a mirage alone. Hence, there is a distinct difference.

- *Selection* is more about the outcome – the results you obtain, what you arrive at after you've leaped over time. In this context (levitation), you're learning to leap over time, which is essential for developing a magnetic ability. By using various techniques, you'll reach a selection of possibilities that you can then piece together. This is where you learn to perform a selection.

- *Prevention* is more about the legal framework – what you can and cannot do according to the rules. Otherwise, we could break time barriers at will.

These are four-time concepts that we may further develop in our future books. Here, we simply wanted to give a glimpse into the possibility of leaping over time. But depending on your particular case, it will determine whether it works or not, as it should only be used for its intended purpose – as an aid during a magnetic journey.

So, what does this have to do with time and space? It has everything to do with time and space. For if you're to learn how to leap over time, which you need to do in order to control it, you first need to understand different time concepts.

– Excuse me, but earlier you mentioned that we should stop leaping over, and now you're saying we should leap over. What do you mean by that? I ask Fabrizio.

– In this case, within the rhythm of time, we need to learn to leap over, but here it's just a part, a time cycle, when we're developing a whole mechanism. A small piece of the puzzle. When humans leap over, you sometimes skip the entire project. You leap over the entire space within time.

This is mirage technique for us. It's also the one you'll use in the future, designed to help you learn to distinguish between what is and isn't real. But before you begin using mirage technique, you first need to understand what time limitation means.

– Why? Can't we learn about it after time and space?

– No, to learn about time and space, you also need to study the differences between the layers that exist, such as within a time limitation. And this is where you'll find yourselves when you land in time and space to learn about the intermediary layer that exists there. There's a gap, and you haven't learned about it yet because you've tried to cheat. You're skipping over the most essential parts.

– What does this have to do with mirage technique?

– By understanding how an intermediary layer works, the one

present in all spaces of time, you'll get closer to understanding the concept of time. For this is what you need to grasp first before you can insert a time device that can limit or shift time.

We will explain, beginning with time limitation.

TIME LIMITATION

Archangel Michael explains:

How, then, can we apply time in a controlling mechanism so that we have enough time to set up and act on what we need to do? We do this by setting a limitation within time, specifically the one that occurs between time and space. There's a limited space between the layers that affect all dimensions of time, allowing you to alter time. Without this built-in time limitation, we wouldn't be able to stop, create, or advance the time aggregates that you will use in the future.

To set time within the aspect of magnetism, where we transform time into a sequence of events, we can intervene and influence it. That's exactly what we do when we introduce a time limitation – we influence an event. This event technique is something you'll learn in the future – how to use a time mechanism to influence and change an event so you can embrace levitation.

– How can we use a time limitation in our future vehicles? I ask Michael.

– It will be used in the same way you currently use brakes and acceleration. There's already a built-in time-limited stop function when you need to halt, and so on, which exists within a time-limited space. But to create a time-limited space, you must first clarify within which branch you want to operate. This must be done before we introduce a time limitation – ensuring it occurs within a certain framework, within a certain space. However, it won't function in the same way as a time delay or time shift.

– Aren't time delay, time shift, and time limitation the same thing?

– No. Let us clarify:

Time delay is used, for example, with a safe where you set a time, like 1.5 minutes, before the safe locks itself. It happens within a specific time.

Time shift is used, for instance, when you record a movie, but you can't watch it until later in the evening. You shift time within an unspecified period.

Time limitation confines a fixed space. You can alter the time here as well, but it neither locks nor changes after a certain time. If you alter the time, then you delay it. Here, we limit the time space instead of delaying or shifting time in the process. So, a time limitation requires a space where time can reside that is fixed.

We can also introduce a block in that same space so that time cannot expand. This allows us to impose a limitation on time without shifting it.

We can also create a larger space, which we do when we magnetize larger objects, such as a vehicle. This is where we reach when it comes to time limitation in magnetism. The larger the space, the more *frameworks* you have to limit time within.

Additionally, we need new tools, and what we're thinking of is a program where we can time-limit a space. We can't set the time in our new program without first understanding what we need. Therefore, we need to introduce a time limitation before creating a program that will manage different events, different units. It will be a program focused on self-governance.

However, we're not there yet, so we're only touching on the basics of what's to come. But keep highlighting the magnetic capability, as there are those who have already begun exploring this area. Listen and learn when they release this self-governing program, as you will eventually develop it so that it also works in your self-driving cars. And once you've adopted this self-governing program, you can also set different time regulations, different time parameters. This is what will determine when a time limitation should occur.

To time something in the future, we also need to know when and at which station this will happen. Therefore, you will also present different proposals on how this development should occur. It won't be an easy task at first because there are those who don't want you to get that far. But we will help you with this, so don't worry.

– What stations are these?

– There will be different time zones, and this is where problems will arise because you're in different time zones where the clock shows different times depending on where you live. But you'll solve this by adopting a common time that everyone can adapt to, and it will be essential to do this before you set your limitations. We will also touch on this when we discuss switching stations. It will vary among different entrepreneurs, but overall, you'll be able to adapt to the same time frame regarding magnetic development.

That's why it's so important that you first study all the spaces of time, meaning the resistance that exists and the movements that occur when time moves. Without movement or resistance, you can't impose a time limitation because it's within the *resistance of movement* that you establish a time limitation. So, if you explore the resistance and the movements that occur when time moves, you'll also find the answers to where to place a time limitation.

However, to control time, it must first be timed, for example, where or at what pace a movement should occur. It's when you discover how time can be controlled that you can also limit it.

Some closing words:

To adopt a time limitation within a magnetic capability, you also need to study the fields of knowledge from the past. If you do, you'll combine the knowledge of the past and present, and only then will you learn how a time limitation truly works, including its application (hint). This was something the Kathremes touched upon when discussing your future remote control. It will be a program that will be integrated into your remote-control system.

TIME-LIMITED USAGE

Angel Fabrizio speaks:

In addition to time limitation, there is also a concept we call *limited-time usage*. By limited-time usage, we mean the period from when you start the car to when you turn it off. When the car is turned off, it no longer receives kinetic energy, and when it's no longer moving, it enters a sleep mode. This will be crucial in the future as well, as you will need to program limited-time usage into your new system.

This is also where you will learn to use the technique of the star, the one we previously discussed. It involves the star itself pausing to enter a conservation mode, then starting everything back up again. Therefore, it's valuable for you to study how the star can halt its process and enter a conservation mode, and what happens internally during that time. Because everything happens within – it's in the internal processes.

– How do we manage this conservation mode? Or rather, can we even do that? I ask Fabrizio.

– It's great that you ask because that's exactly what we can do. This is what we touched upon earlier when we talked about managing magnetic kinetic energy. This is what happens in a levitation car. And to understand that, we need to understand how a magnetic process works.

A magnetic process always generates energy particles – particles that are transferred between one another. When these particles are no longer in use, we can manage them, meaning when they're in saving mode, such as overnight. This way, just like the star, they can restart everything after a period of rest. It's essential that this type of magnetism gets to rest sometimes because the levitation car, at least initially, will only function for a certain period. But later on, it will be able to run continuously without needing to pause.

Let's take a train as an example:

According to this method, the train itself, and only for a few seconds at each stop, will enter conservation mode. It's at this point that more particles are added, and when that happens, they increase in number and strength when the train is set in motion again. And when you've learned to manage magnetic components (particles) in this way, the particles will only need a few seconds to recharge and restart everything.

But to get a coupling system going, we also need bolts. For us, bolts are an integrated concept, where we bolt together the parts we want to integrate with. It's not a physical object but a term for the process that needs to happen. Therefore, we also need knowledge of how to manage energy to start and bolt everything together. This is what will be important. At each stop, the information stored in your coupling system will remain. You can reactivate it when needed. We can call it a reserve tank.

This will be a significant lesson for you in the future. That's why we wish for you to revisit all the projects related to magnetism that were previously shelved. Our advice is to start searching through old engines. There are many documents there that describe how to properly manage a magnetic kinetic component to save energy. They're in old, brown, and worn-out desks. So dive into this! You'll find the foundation of this knowledge, and from the notes you find, you can further develop the capabilities of magnetism. Some researchers have already begun to understand this concept. If you read our words, we're happy to assist with this matter.

This is what mirage means to us – connecting different levels of dimension. Understanding illusions so you can put selection into practice. Prevention won't come into play until you need to implement various regulations in the future.

TIME CONTROL UNIT

Angel Fabrizio continues:

To connect everything together, we also need a time control unit. A time control unit is essentially a device that governs time – it's what controls time. This could be anything from a time machine to a train, just to give a few examples. Our advice is to continue with trains, as you've already started working with them.

A time control unit is the part that manages time. Unlike the timeline, which we'll discuss later as a fixed entity, a time control unit is an element that interacts with time, such as in a car, so it knows when and at what speed a transformation of particles should occur. Therefore, a time control unit is a crucial component, but the specific time it operates on is set within a computer program.

However, before you can do that, you first need to learn how to pattern-match, and this is where the solar pattern we mentioned earlier comes into play. Once you've learned to pattern-match, you'll also learn to adjust the speed – both in terms of when and at what speed a transformation of particles should occur. This includes how time intervals should be integrated. This is part of what happens in a pattern-matched solar system.

When you've mastered pattern-matching, you'll know when different processes occur and at what stage a transformation is in. Then, you can program everything into your future time control unit.

– So it's like an extra machine where we input various time specifications for when a transformation should occur? I ask Fabrizio.

– Yes, but not entirely. This isn't a standalone process or part. It integrates with other time specifications, such as when a start or stop will take place. The difference here is in determining what should be transformed and the strength of that transformation – within the realm of time. That's the main purpose of a time control unit in this context. Later on, you'll develop your time control unit further so that it can synchronize with other time specifications and apply to other tasks as well. But start with what's in the transformer and learn from that.

To understand how a movement sequence works, we first need to understand how time and space function. This is what will determine whether you can achieve a more advanced concept of levitation. It doesn't matter how skilled you are at developing magnetic abilities – if you don't understand time and space, you won't be able to learn how to control time.

Once you've done that, you can learn to connect a time control unit with the magnetic capabilities of the future, which you can then integrate into any machine you choose. And to develop a levitation car, you first need to learn how to connect. Because it's by connecting components and understanding how time and space work that it will become easier when you eventually apply magnetic capabilities.

MAGNETIC TELEPORTATION

Kazandra speaks:

When you can handle the smallest components, you will also learn how to teleport a magnetic process. This is what you need to understand before you can fully manage the magnetism of the future.

Teleportation involves transferring information from one place to another, and in this case, we are transferring information from the past and future into the present. It's like a kind of intersection between the past, future, and present, and what happens between these times. This is what we want to explain now. Of course, there are other concepts to explore, but we will start with this.

A magnetic teleportation is an event that occurs between time and space. Therefore, it is crucial that you first understand how time flows between spaces. This is exactly what a magnetic teleportation does – it moves and changes between time and space. Once you understand how time and space work, you will also learn how to demagnetize a process. Demagnetization is about stopping the magnetism in the middle of an event because sometimes that's necessary. So we stop a magnetic ability before we can demagnetize it.

– Isn't stopping and demagnetizing the same thing? It sounds like it, I ask Kazandra.

– No, demagnetization is performed to interrupt an event in the middle of a process, but everything around it continues. We strip it down so it can no longer produce, but not so much that everything around it stops. When we stop, we stop everything.

So, what are we going to demagnetize? The part that we, Kathremes, want you to start with is the magnetic ability, which you do through teleportation. This is where the next process comes in. By teleporting, you can enter the realms of the future and past and change the concept, or rather, the content. This is what we do when we demagnetize a magnetic ability. We do this to alter the content. But to do this, you first need to create a kind of stop where the magnetism ceases to work, so it stands still. Everything must be completely still in the process you use to teleport an event.

By using teleportation, you can also ensure that everything is going as it should in time. When we teleport, we can ensure an event and add or remove particles that should not be there at that moment. This means that they are either too strong or too weak, which we discussed earlier. This is what we can observe and do in a teleported magnetism. We can, in advance or in the past, see what has happened or what will happen and act based on the results we get.

– How can we change something if it's in the past, if the particles have already weakened and need to be burned away, since otherwise they lose their power?

– Oh, what great teachers we've been (laughs). It's wonderful that you understand this now. You see, everything you have learned so far will lead you to teleportation, which can enter the particle flow of the past and future to ensure that what happens can always be shown. Because as it stands now, what has happened, has happened, and we can never go back in time since we can never turn back the clock. But in a magnetic teleportation, we can do that and, as you said, burn away the link, the particles, that have not transformed as they should. That's why we teleport. We can say it becomes an extra tool where we ensure that the magnetic process goes as it should in all future spaces.

– How do we teleport magnetism?

– We always do this by stopping an event, or as we previously mentioned when a star rests, so that we can change a pattern. This is done using a secure method we employ. Then we can retrieve past realms and insert them into the future or present. It might seem a bit complicated, but you will understand once you're there. There is also an intermediate layer in this magnetic teleportation. That's why it's so important to understand it first before you get here.

Additionally, there is also a *marker*, as we call it. It's a line that shows how, but mainly *when* you can teleport a magnetic concept. But we'll cover that later.

– So we teleport to secure a movement, in this case, a magnetic movement? And we teleport the magnetic process to ensure that everything works as it should?

– Yes, exactly.

– When will we get to the electromagnetic concept?

– We'll get to that in the next phase, which we can't discuss right now. We'll cover it once we've settled in with you because that teleportation process must not fall into the wrong hands.

– Thank you, I understand. How come we can use teleportation within the magnetic concept?

– It comes, like many things, from an event where a planet could retrieve completely different information from another time. It was also after this that they could further develop this concept to work in other areas beyond just informational ones.

– How did they do that, and can you tell us who they were?

– We can, they came from the Pleiades realm, where the ancient Pleiadians focused on how they could change an entire concept based on the patterns of the past and future, which they called information of the past. But they did not live where they live today; they were predecessors to the present Pleiades realm, so to speak. They were also a very positive people who traveled a lot to develop various methods, as they said at that time. That's how all traveling began, which we now call teleportation. Because that's what happens when we retrieve events from the past and future – we travel through time and space.

So start with magnetic teleportation, and we will further develop the electromagnetic aspect for you in a few years.

– Surely, we can teleport much more than that?

– Yes, indeed, and that's what you will develop further down the line. For example, we can also teleport cells, those in your body.

– That sounds dangerous.

– No, not at all. It's perfectly safe, and this method is used to ensure, repair, change, and so on. But this won't happen until further down the line, as you first need to increase your inner strength, but that will come.

CHAPTER 19

INTERMEDIATE TIME – INTERMEDIATE LAYER

COURSE OF EVENTS

Kazandra speaks:

Once you understand how time and space works, that time can be controlled, and you've learned to manage different time regulations, we move on to the actual course of events. It is a process where you can pause and amplify the flow of time. This is a technique we already use on our planet. You will also develop and use it in your magnetic abilities.

So what do we mean by the course of events? When an event occurs, it usually happens within a certain time period – it begins and ends within a specific timeframe. This is a course of events. By understanding this, we can also intervene and stop an event happening within a certain time by interrupting the event itself.

– How can we break an event? I ask Kazandra.

– We stop the event before time runs out. But to stop an event, it must first have started. Otherwise, we can't stop it. When we proceed to break a course of events, we always begin by dividing the entire process into different epochs, which helps us identify where in the event we want to break time. It is through the knowledge of how to interrupt a course of events in time that we can learn when and where magnetism should be disrupted.

After dividing time into different epochs, you start by installing a heat-generating unit. Initially, movement will manifest as heat – before you achieve complete levitation. So at first, it will be a heat generator where your unit works with the various epochs of time. After that, you will shift the time rhythm. The epoch that applies when you start using levitation is entirely different from when you use heat. We know it may seem confusing, but your scientists will understand.

This is the first part of a course of events we want you to further develop now. You will also add more events in the future that will belong to the same epoch. When you have learned to control time, you can also learn to unify it. And once you have mastered unifying time, different events can occur within the same timeframe. This is how we utilize our magnetic ability, which we use to control in which direction we want the magnetism to pull and within what timeframe this should happen.

For example: If we want the magnetic ability to act forward between 2:00 PM and 5:00 PM, we need to set that time. Then, if we want a stop at 3:00 PM, we can use our controlling unit to stop an event, an epoch, to temporarily pause time. Then the journey continues to other stops, and so on.

Think of a driver stopping at every bus stop. But here, we're talking about a self-thinking magnetic ability, whose time and rhythm can be set depending on what needs to happen that day.

We also need to introduce different gates. You will pass through gates when calculating time, as these are the gates that affect time. They are gates that act upon time. Therefore, it is the time, integrated into various gates, that we need to calculate.

- A gate is like different stations where the train stops either to drop off or pick up passengers. And it is at these gates that you will learn to calculate time. Without these gates, we get neither gaps nor layers, which we need in a magnetic ability.

- A gap is the transport stretch from one gate to another.

- An intermediate layer is the time it takes from when the train has stopped until it starts again. It acts as a dividing line. And it is there, within this layer, that the gates are located. It is there that a change occurs, where they recharge and add more power. It is also fundamental that you allow your future magnetism to have a layer where we can calculate and have room to maneuver.

– Isn't a gate and an intermediate layer the same thing?

– To some extent, but not entirely. A gate ties together what you need to calculate within an intermediate layer. So both can be integrated into an intermediate layer, but they still represent different factors. Do you understand?

– No, not really.

– Let's try again.

A layer is where we *perform* the actual calculation. A gate is more of an event within the intermediate layer where you *make* the calculation. A gate opens up for the calculation you need to make *within* an intermediate layer. Is that clearer?

– Yes, thanks. Now I understand. So I can calculate what is in a layer using a gate?

– Yes, exactly, but with a small twist. The calculation that occurs within the layer can never change. It is permanent because the space is also permanent. But the gate within the intermediate layer can vary because it is where you make the actual calculation between time and space. So you can make a new calculation, but once it is set in the intermediate layer, it is permanent. Therefore, they are alike and integrated into the same space, and it is in the gate that you can calculate the smallest – within what happens between time and space.

COUPLING STATION

Kazandra speaks:

You also need to study various coupling stations because they will control your future remote controls. A coupling station is the intermediate layer where we switch between different time zones so that we can use time to either move forward or backward, depending on the time setting you input.

– How can time go backwards? I ask Kazandra.

– The idea of time going backwards in this context is more about the time setting you input. So it's not time itself moving backwards; rather, you will learn how to reverse time within the program you are using. This is necessary to unify your different time zones before reaching a common time zone. Sometimes, it's also necessary to reverse time so that you can review what went wrong. It can also be used to clarify if a time setting is misplaced within its own space.

– So initially, the timing in a coupling schedule will differ, but what happens after that?

– Then you integrate everything into a common case.

– What is an intermediate layer in this context? Is it the same as before?

– A layer in this context is more about, as we mentioned earlier, taking a break if you need one. Think of it like when you're out jogging or working and need a break halfway through. This, too, is an intermediate layer.

– So there's an intermediate layer in every aspect of time, but not a gap? What's the difference?

– To us, a gap is when you completely skip over an event. An intermediate layer is where you pause, either like the jogger catching a breath or like a train switching tracks. But you can never skip over an event in time, which some researchers believe you can, but that's not the case. What we're talking about always involves an intermediate layer where you change an event to move forward. And it's always within an event, in a layer, that we make connections, never before or after. If someone says otherwise, it's wrong.

– Why can't we connect things before or after an intermediate layer?

– Because another process is happening at that time, and if we try to connect everything with what's going on then, it creates what we call a *cross-connection*. It just doesn't align. That's why all connections are always made within an intermediate layer, and a coupling station won't work without an intermediate layer. So for a coupling station to function, we need something that can integrate into an intermediate layer since that's where all connections are made.

– Aren't a coupling station and an intermediate layer the same thing?

– No, an intermediate layer is just a space. A coupling station is the part that acts on time within the intermediate layer.

– What does a coupling station do, and how many can we have in each system?

– We can't have more than what an integrated coupling system can manage, as we can't manage an unlimited number of time settings, at least not initially. So each coupling station in the intermediate layer has its own time setting. As mentioned, every intermediate layer is where a start or stop happens, and that's where the connection needs to be, the one that switches between different times.

– Wasn't that the job of a time shifter?

– True, but that's not what we mean here. In this context, we calculate when and where a train needs to stop or start. So a coupling station represents more precision, with everything set as a specific time setting in a computer program. A time shifter only functions as an additional component but doesn't have a decisive role. So they are different, but both are still affected by all aspects of time (we will discuss time shifters later on).

– Why is the switching station so important? And why can't we just have one consistent movement in time?

– If we're always affected by the same thing, we don't evolve. So if we don't have a dynamic coupling station, no development occurs. The coupling station is also there for the future, as you will also learn to transmit information, meaning that coupling stations will begin to communicate with each other. But you haven't reached that point yet, but you will explore it later on in the future.

That's why we wish for you to start developing a coupling station that is dynamic. It is important that each coupling station can be changed; otherwise, they cannot integrate with each other, which is our goal. So study and develop an integrated coupling system, and you'll see how easy it actually is when you later need to unify everything.

Fredzo adds:

It's also important for you to study more time settings to see what works in each coupling station. Each coupling will represent its own part, which is why they can act in different ways. This is how you get different readings when you combine time with a start or stop.

It is also our wish that you do not use too many resources at first. You might think that each project requires a large construction and high costs, but that's not the case, as the costs are already set. So if someone says it's going to be an expensive project, it's not true. You can build with very small means. That way, every citizen will be able to afford a more environmentally friendly vehicle in the future.

INTERMEDIATE TIME

Archangel Michael speaks:

We also need an intermediate time within the intermediate layer. This will determine when time should move forward, when the cogwheel should turn (we will discuss the cogwheel later).

– How can there be an intermediate time within an intermediate layer? I ask Archangel Michael.

– The fact that we can have an intermediate time in an intermediate layer means we can divide time into different zones. This is something we mentioned earlier when we talked about achieving a common time zone. Therefore, it's important that you also implement various intermediate times.

– How does an intermediate time work, and what is it in this context?

– An intermediate time is a shared concept, as we often say. This is because it's during this intermediate time that we can adjust time. We can't change or move time within an intermediate layer without an intermediate time.

An intermediate time sets the tone for time and indicates when and where time is located. Therefore, we need to know where time is before we can set it, and this is where the intermediate time comes in. The intermediate time determines when and how a time indication should occur, including when a stop or start should happen. That's why we need an intermediate time within the intermediate layer. We are always in an intermediate layer between different time indications, and in this case, it's an intermediate time.

– So it acts as a controlling factor, but only in the part where we can set the time?

– Yes, exactly. We can't control time without an intermediate time, the one that exists between a start and a stop. Therefore, you need to arrange all the times you need to develop. Then, you insert an intermediate time into each event, to finally establish a common time zone. So, it's essential that you reach a common time.

– So an intermediate layer is like a space in the gate where time can change, and it's the intermediate time that controls everything. Without an intermediate time, we can't set a time indication. Did I understand that correctly?

– Yes, exactly right, but with a small twist. For time to function within an intermediate layer, we need something that can act as movement. Otherwise, the cogwheel can't move according to the different aspects of time, and this is where we introduce a start and stop in an intermediate layer. So, the intermediate time ensures that everything works as it should, that your time indications function properly. And having a layer where an intermediate time can operate is crucial; otherwise, we can't set time to a unified internal concept.

An intermediate time is what happens in
an intermediate layer
where time is governed and controlled

– How do we reach an intermediate time?

– We don't reach an intermediate time. The intermediate time is a set time that you determine, and it's the one that the cogwheel then adjusts to. For each step a cogwheel takes, there's a new intermediate time in the intermediate layer.

– So we can have several different intermediate times in the intermediate layer that the cogwheel moves according to?

– No, not like that. We can only have one intermediate time in an intermediate layer. That's why we need different intermediate layers controlled by the same intermediate time, with different time indications placed within that specific layer. As a result, the intermediate layer gradually shrinks as more events are added to it. It doesn't physically shrink; it's just a feeling we get when something fills up more and more.

Think of it this way, and let's use a train as an example:

The train stops at a station; this is an intermediate layer. But for the train to know when to start again, it needs a time that tells it when, and this is where we insert an intermediate time. A time within time, you could say. When the train starts moving again, it's in a transit period until it reaches the next intermediate layer. There, another intermediate layer takes over and controls when the train should start and stop at that station.

But to prevent an intermediate layer from being affected by multiple intermediate times and disrupting the rhythm of time, we insert a time indication that is adjusted so that the time switch and cogwheel follow only that specific intermediate time, at that specific location, at that specific train station. In this way, we give it a command in time, so to speak, so that it can sense when a start or stop should happen in each intermediate layer. Do you understand, my friend?

– Thank you. Now I understand. But what about the transit period? How does it work between each intermediate layer?

– It acts as a driver, we can say. It's where you determine the speed so that each transit period takes exactly the same amount of time before reaching the next intermediate layer.

– How do we know how long each period should take?

– Your professors will calculate that, just as they do today. But the same transit period can't be calculated the same way each time. Every time a train runs the same route, it can't know exactly when a start and stop should occur. That's why your future stations will calculate this themselves when they arrive at the next intermediate layer.

– How can they do that?

– There will be an event program where every event along the same transit route is recorded. From this, the train, or rather the program, can determine when it arrives at the next intermediate layer. But this is a more advanced line, rather than a transit period. However, we want to include this now so that those who have already started this significant work and are not believed will know that they are on the right track.

Everything will eventually be self-controlled in the future, even the obstacles a train might encounter along each line, during each transit period. The train itself will intervene and adjust the time so that people know when the next train arrives at their specific station. So it will be a time-scheduled system that you will follow, but the program itself will adjust the time if something unpredictable happens between each intermediate layer. That's why we need an intermediate time in the intermediate layer.

You will also add more components where each component controls time. We've broken it down into three points:

1. The first is different time zones. This will include a program that incorporates all the time zones you have today and can set itself to a common time. These are different components that you need to add before reaching a common time. This is your first component.

2. The next component is different cycles. Initially, you need different cycles that can determine speed. It is the speed that will determine when a train should start or stop at each intermediate layer. This means you need to achieve a common speed so that you can calculate the transit period and determine when each start and stop should occur.

3. The next part is a transformative program. It's a program that can independently set the transformation that should occur in the transformer. However, it first needs to reach another power source that can be transformed before becoming more self-sustaining. So, it will be a power source that you will use to help you before achieving a self-sustaining program for particle transformation.

Start with these three components: *zones*, *cycles*, and *programs* before developing a more self-sustaining concept. Your professors will understand.

COUPLING CONTROL

Kazandra speaks:

A coupling station is also part of what we connect to a control system, including remote control. Therefore, we also need a control switch. A control switch is what governs the entire project, and it's a connection that needs to occur within the intermediate layer we previously mentioned.

It can also ensure that there are no other obstacles in the way within the intermediate layer. If something else is in the way, it could result in a faulty connection. However, in this case, the mistake isn't about connecting in the wrong place, but rather a misconnection with the wrong control, hence the name. It's precisely in the intermediate layer where the control exists, and where we connect it to our remote control.

– What kind of control is there in the intermediate layer? I ask Kazandra.

– It's a control that can operate through remote signaling. It sends signals to you when you've entered an intermediate layer, and that's when your connection takes over. It will be governed by a schedule that your integrated switching system follows. You can refer back to our image, where you can see what we mean by an intermediate layer in an integrated switching system. Within that layer, there will be a control and a time that you can adjust to, depending on the time indications you've programmed. So, to connect everything, we need a control in the intermediate layer. This control then links everything so that time can continue its journey to the next intermediate layer.

– Ah, I understand now. It's like a movement schedule, where each stop and start form an intermediate layer, and depending on the time indication that's been programmed, the control operates according to the different dimensions of time. But isn't the control affected by the surrounding space?

– No, in this case, the control switch is independent because it's more influenced by an internal concept – what you input into your data. Think of it like a railroad crossing and a switch. The train stops when the barrier comes down and starts again when it lifts. An intermediate layer, in this case, is like the switch itself, which is needed when the train needs to change direction or track. So the switch shifts to another track before the train arrives, which can be compared to a control switch in an intermediate layer. And each switch is already pre-programmed in a data system. That's why we also need a time switch.

CHAPTER 20

CONNECTION TO TIME

TIME SWITCH

Alfredo speaks:

To link everything together in an integrated coupling system, including the power-carrying chain we discussed earlier, we also need a universal time switch. We'll draw up a simplified diagram of a time switch, and we've broken it down into three parts:

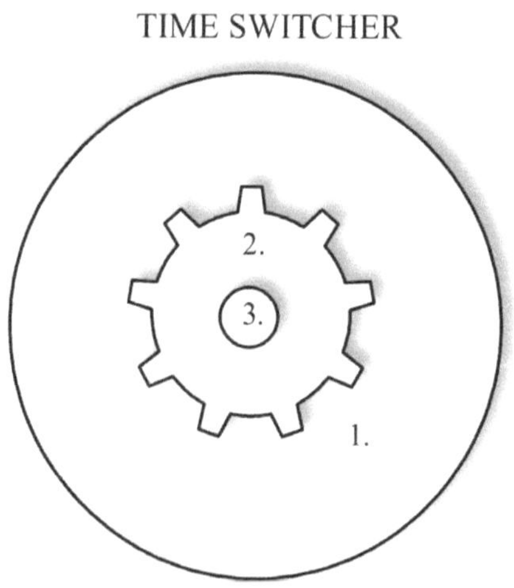

1. Space = Time determiner.

2. Cogwheel = Moves according to timing rules.

3. Nozzle = Acts as a power source from inside out.
   The nozzle draws energy from the power-carrying chain and supplies the force needed to rotate the cogwheel. The timing entries dictate when the time switch should stop or start.

- The time switch determines when a start and stop should occur.

- It's a time switch that you connect to your pre-installed data program.

- It controls time within an integrated coupling system.

- The time switch must be understood from a magnetic perspective.

A time switch is something that, together with your computerized program, will control time. Therefore, we first need to understand how time and space work, as this is where everything is created, built, and developed in the progression of time within designated spaces. A time switch is thus connected between an integrated switching system and a data program. You need to have developed an integrated data program before you can reach this point. Once you have done that, you will understand what we mean. Therefore, we are not developing the data program here; that will come later.

For now, we just want to highlight that a time switch will need to be connected with an integrated coupling system.

– How do we do that? I ask Alfredo.

– You do it by creating a coupling system that can transition between different time zones. You will develop this time switch globally and in various contexts. Your professors, who work in this field, will create it for you.

– Can you explain how it works, for those of us who don't have much knowledge about this?

– We can put it this way: The time switch you will create in the future will not look like it does today; the current one is more visible to the eye. In the future's time space, it will be created in an earlier designated space that only the most skilled professionals can see with the naked eye. So, it exists but is still abstract in its content.

By abstract content, we mean the effect that time has when you create a time switch. Time provides an abstract content, which we then lock away, so to speak. That's how our time switch works on our ship, and only our mechanic Fredzo knows more about this. So, we'll let Fredzo take over now.

Fredzo explains:

To fully understand time and space, we need a container that can self-regulate according to your data program. It's a time content that consists of a *clockwork*, which we call it. Think of it as the contents of a clock, but this is a more advanced and determining technology.

– How can we connect and understand how this abstract content works? I ask Fredzo.

– The content will function almost like an integrated coupling system; in other words, we need some form of force that makes the content rotate. Otherwise, the clock cannot rotate either. So, to make it rotate, we need an instrument to do that, and in this case, we're talking about a nozzle.

– What does a nozzle do?

– A nozzle is placed at different stations where we need to introduce a time stop or start. The nozzle controls that part.

-But where does the force come from so that everything rotates?

-It comes from a cogwheel. The cogwheel operates according to the nozzle's determining part, which is already set. This was what we discussed earlier regarding the cogwheel's movement over time. The force is created externally; it comprises the entire force that governs the integrated switching system.

– But where does the force come from so that everything rotates?

– It comes from a cogwheel. The cogwheel operates according to the nozzle's determining part, which is already set. This was what we discussed earlier regarding the cogwheel's movement over time. The force is created externally; it comprises the entire force that governs the integrated coupling system.

– So the force is introduced into the time switch, the container?

– No, it enters automatically because it links with other concepts in an integrated coupling system. In this case, from the outside to the inside.

– So we have a container with a nozzle that integrates with the external force present. The cogwheel is just a factor that affects whether time moves or not. Is that correct?

– Yes, exactly. The force will eventually merge. But you're right; it's also a self-operating mechanism because, as we mentioned earlier, it's a time switch that can be integrated into other matters as well.

– But time, how is it determined?

– Here is where your data program comes in. You input the time yourself, wherever you want. Then the cogwheel moves according to the movement principle, and the nozzle always sets the pace since it also has a predefined determining concept.

– So we need to learn when to switch between times. And can the cogwheel have irregular cogs for time to occur differently, or are they uniform as in the picture?

– A cogwheel is always uniform. It's the time in the space that determines or decides how or when everything will happen. By time, we mean the period of space between the cogwheel and the space where time is situated.

Think of it like a game where each move brings everything closer to the future because time always moves forward. Or consider your own life: sometimes you need to pause (rest), sometimes you need to be completely still (sleep), sometimes you need to start everything again the next day (wake up), and then it all continues. Therefore, it's crucial that you choose the right aggregate for the time switch; otherwise, it won't work.

Consider this: Time always moves forward, but since we need to stand still sometimes, we also include that in the time switch. This ties into what your angels discussed earlier; for example, when the train stops at 15:00, we need to include that in the time switch. This is where everything will interact.

– How does the time switch know when the train should stop?

– You input it into a data program that synchronizes with the cogwheel. But it's in the actual space, what exists between time and the cogwheel, where everything is determined.

– I don't completely understand. What do you mean?

– Let's use your kitchen as an example, where you're sitting and writing with me right now.

The kitchen is your space, but time flows around you. However, you can't stop time yourself. To do that, you need an integrated program that can do it for you. So when we talk about placing time in a space, we mean placing the time input within the space. You can never place a time change inside the cogwheel. The cogwheel is controlled by external force, so we also need to place the time input there.

Therefore, all time is placed and changed within the actual space where you work. Never on the inside. If you place the time input inside, for example, the nozzle, it won't reach what happens outside, where the force is controlled and present. The risk is that it remains stuck there.

– But in a computer, we place the time input on the inside.

– Yes, but here your computer acts as an external connection. It's not where the actual time inputs occur, even though the time indications are entered there. A time input always happens in a space where the force is accessed from the outside. So, it's the time in the space that's important in a time switch. The cogwheel only acts based on the internal force it gets from the nozzle and stops or starts according to the time inputs you provide.

TIME COUPLING

Alfredo speaks:

You will also need a time connector, and the time connector we want you to study represents an internal concept of time. Therefore, it's crucial to first study all the dimensions of time before you integrate a time connector into a time switch. A time connector is indeed a component you will need in your future time switch. When time comes to a standstill, the time connector will take over to ensure that time never stops.

– So it's like a security guard? I ask Alfredo.

– Yes, you could say that. What we mean is that the time switch needs to be equipped with a time connector that can detect when time is going wrong. Nothing is perfect, not even electromagnetism. Since time can always be altered, there's also a risk that something might go wrong. So, a time connector is simply there to ensure that time in a time switch does not come to a halt.

– How do we integrate it with time in our future time switch?

– Your engineers will handle that, but it won't happen until you understand all the dimensions of time. So our advice is to start with that.

TIME AXIS

Kazandra speaks:

To unite our time aggregate in the future, we will also need a time axis. This is what determines the direction in which time will flow. By axis, we mean the element or thing that acts for time.

– How does it do that? Is it similar to the cogwheel? I ask Kazandra.

– No, a cogwheel always needs a space where time can unfold, whereas a time axis is a consolidated history. Therefore, it is not governed by time. It does not control time; rather, it only operates based on what you input into your data program. It cannot act on its own.

– What then distinguishes a cogwheel from a time axis? To me, they sound almost the same.

– We understand it might seem that way, but in a magnetic process, a cogwheel acts by moving forward, while an axis merely exists and is fixed in its own function. Do you understand, my friend?

– I think so. But how can it act if it is fixed?

– It is not completely fixed or stationary. In this case, a time axis acts according to what happens around it, not like a cogwheel that turns with time. An axis is simply a part of time's functioning.

We know that for you, a time axis is a chronologically ordered list of events – what happens over time. But in our case, it functions a bit differently, though not entirely. If you understand? Look at point five in the image. Then you'll see the analogy between a chronologically ordered list of events in time and our time axis. Hence, we refer to it as a part of all spaces in time.

– Where do we place it?

– You will place it together with your future transformer. It will be important that your time axis can operate based on what the transformer emits. Thus, the time axis acts according to the force created by the transformer, but it still remains as an individual entity in its own actions. It can also act according to other commitments. In summary, an axis can act, but it can also remain independent. It does not move, but it responds to the force it encounters, in this case, the transformer.

– So it acts according to the force. Why does it do that? I mean, what is the purpose of that?

– By having a time axis on the outside of the transformer, but still linked with all the content, you can also see what force the internal component emits.

– Isn't that the temperature sensor's job?

– Both yes and no, as it does not pertain to the aspect of time. It is essential that the force affecting time in the space can integrate with both the external and internal aspects. This is what the time axis will do for you, which a temperature sensor cannot do since it does not exist within time; it merely measures the force. By first measuring the force, it becomes easier to capture the strength that should pass through time's space.

So, you see, a time axis exists there and acts as a sensor, reacting to what it detects, but only within the space of time. Initially, we measure the force, which then passes through the time axis that acts according to the force in all spaces of time. If you don't understand everything yet, it is because you haven't learned about time and space yet. But when you do, you'll say, "Ah, so that's what they meant." So don't worry if you don't grasp everything we say right now.

TIME CAPSULE

Alfredo speaks:

We also need a time capsule. By time capsule, we mean how everything can integrate. It's like something or someone that ensures everything goes as it should, so to speak. A time capsule is the component that can integrate with almost everything in a transformer. That's why it works so well with almost everything we manufacture.

– How can a time capsule function in and with a transformer? I ask Alfredo.

– It doesn't work within the system itself. It's a component that exists on the outside but operates based on how everything synchronizes on the inside. A time capsule is placed on the outside. It ensures everything goes as it should. It acts, decides, and makes sure everything functions according to all spaces of time.

– I've learned that a time capsule is a container where we place other items to be preserved for the future and then bury it.

– We understand, but that's not how it is for us. Our capsule is a part of time. Without a component that can sense, understand, and control, many errors would occur. So a time capsule detects if something goes wrong inside.

– Isn't that the job of the temperature sensor or the time regulator?

– No. A time capsule "knows," a temperature sensor "understands," and a time regulator is an "internal" process, so to speak. Therefore, the time capsule is a more decisive factor than, for example, the temperature sensor.

The power that connects the transformer's inner workings with the outside is the one that provides the energy the time capsule senses. It can detect if a part of the transformer isn't functioning correctly in the space of time, whether it needs to be replaced or repaired. In other words, it is a capsule that is determined and controlled by time. Hence, it can act and sense whether the internal components are working as they should.

– So it functions like a kind of sensor?

– Yes, exactly. It senses the rhythm of time and alerts you if any part isn't working as it should. Thus, time controls, but power sets everything in motion. Think about what your angels discussed earlier – without time, we cannot have forward movement. This is what our time capsule is for – to work. Therefore, it is a controlling and decisive factor in a transformer.

– Where should we place it?

– You can place it wherever you like, but our advice is the right corner or one of the other corners, as that's where it senses the strength best. You'll understand, so don't worry.

Now, my friends, we've created a transformer for future use, so that you can eventually develop a levitation vehicle. But before we finish, we need to integrate and test everything.

CHAPTER 21

INTEGRATION

TEST RUN

Kazandra speaks:

Before we began our part, your angels spoke about the power of the universe, because that's where you'll draw pure energy from in the future. After that, we moved on to our part – the magnetic concept. This is where our expertise lies, and it's also what we'll help you with in the future.

We started by discussing the magnetic ability and how electromagnetism is going to get a significant boost in the future. We also talked about the part that has been distorted and how you'll resolve your ascension in the future. Then, we wrapped up everything with time and space to connect our transformer with both a generator and an integrated switching system. The next step is to integrate and test everything.

So, how can we integrate everything? When we Kathremes put everything together, we use a technique that we call *project merging*. What is project merging all about? It's about seeing the big picture and how we can foresee the final outcome right from the start. But also, in what context we will use all the components.

The next part involves how we manage all the components. We want to make sure they can be disassembled, repaired, developed, moved, and used for other components as well. Today, you only manufacture products for a single purpose, like a car. This approach is outdated. You need to further develop each part so that they can also be used in more stations.

After that, we multiply everything – we exaggerate everything. That is, when we assemble all the parts, like remote control, time switcher, and integrated coupling system, we exaggerate the execution of the process to ensure that all the parts are truly compatible with each other. By doing so, we minimize the risk of something going wrong later on.

Finally, we wrap everything up with a test run. Here too, we continue to push and increase the pressure even further, which means we exaggerate and test the power coming in. We also increase the pressure on the cables much more than they can actually handle in terms of heat. That's why parts sometimes break during a test program.

This is how we ensure our products are safe for our citizens. And since we can't sit in the train or car ourselves during the test run, we have machines that handle that part for us, just like you do. Because it's important that we never release a product until it's completely safe.

And don't forget: there will be some puzzles at first, but your scientists will solve them through various meetings where they discuss this integration. So, don't worry. But we can say this: when we were learning to bring everything together, it required a substantial electrical barrier, the one we discussed earlier. This was to prevent one part from knocking out the other, which can happen initially before they've learned each other's functions.

So, our advice is to assemble everything in a way that it can also be integrated into other applications while you're working on integrating different parts. Think big and think forward, even in other areas.

NEW AUTOMOBILE INDUSTRY

Kazandra speaks:

When you are done with the magnetic concept, you will create an entirely new automotive industry. Why focus on cars? Cars will become increasingly common as vehicles as more people can afford to buy them. There are many countries today where people can't afford a car, but as the world changes and you begin to collaborate with each other again, they too will have the pleasure of owning a car. The world is meant for everyone, not just those with the most money.

So, how do we build and construct a new automotive industry? We do it by using a very old concept, one that we also used when we developed our automotive industry.

We've divided it into four sections:

1. Treatment
2. Activation
3. Uniting the old with the new
4. The Power

The first thing you need to do is *treat* what can no longer be used. By "treat," we mean how you can already start transforming today's concepts and industry into those of the future. This is done by using old parts and treating them with an entirely new method. It's a new type of treatment that we will discuss with you in the future, but only once there are no longer any egocentric souls left on Earth. That's why we're waiting – so that everyone can participate in and benefit from this treatment concept.

Through this new technology, you'll be able to convert and use old constructions and objects in a new configuration. That's why we treat them so that old items can also fit into a magnetic concept. This way, we are more environmentally conscious than if we were to manufacture new items, which isn't completely necessary in this case.

Once you've begun treating your vehicles and the parts you want to transfer to a magnetic concept, you'll start phase two. Phase two involves more external influence, even though we're working internally. Phase two is about how you can activate old items so that they can be used in a more future-oriented concept. It's about activating old concepts, including those you discovered during the Roman era – concepts that others have buried so that you wouldn't access that kind of activation.

– How could people during the Roman era have access to activation when today's items didn't exist back then? I ask Kazandra.

– We don't mean it that way. What we mean is a universal technique, a universal activation, that we Kathremes also work with. Back then, it involved how different ingredients could be combined, and they experimented with different lighting functions at that time. That's the activation we're talking about.

So how could the Romans, even back then, understand and know how to activate lighting? This is precisely what you will learn – the same technique they used, but in a more modern version. The activation we're after concerns the stream of activation, which you can apply to the power of light since that's where you will incorporate this future concept. So, activating new light will be important.

After that, we move on to the third part of your new automotive industry, where we dismantle the old to unite it with a new concept. The third part is about *combining the old with the new.* Therefore, you will dismantle the old, repair it, process it, and convert it so that what you want to use can also be utilized in a magnetic concept. So, the unification of old and new will be crucial.

The fourth and final part focuses more on the execution of the concept and understanding where all the *power* comes from. The fourth section will be significant because it's here, in section four, that all the power will be generated. The power in your future motorized parts will be so immense that it will require its own section, even though it will be spread across the world, with everyone working from the same constructed concept

We will also have a long chain of actors working between different countries to ensure that everyone is working towards the same goal, although this won't be necessary in the long run. But initially, it's important until you can fully collaborate on your own.

Leaders from the Pleiadean realm will oversee this part so that you receive the guidance you need. That's why more of us, including us Kathremes, will work with you to help you move forward on the journey we will take together. Our expertise lies within the automotive industry and other motorized concepts, and that's the agreement we have with your God. So, embrace the teachings we are here to offer and learn to work together, not based on who has the biggest wallet.

Also, remember, as we mentioned earlier, when you start manufacturing cars within the magnetic concept, you will need students who step in and help you in that process. It's crucial that you begin to work together as both teachers and students, as this concept will drive your future development. If you do this, you will achieve much greater results. And don't forget: your students carry a lot of knowledge within their souls. So, start seeing everyone as equals.

CHAPTER 22

CONCLUSION

ELECTRIC CAR AND HYDROGEN CAR

Archangel Michael speaks:

In conclusion, we want to answer a question that we received from Ulrika, and that question is: Why don't you talk about electric cars or hydrogen cars? The answer to that question is simple: According to our source, the hydrogen car is not yet fully equipped. It functions by today's standards, but we believe it is not fully developed.

The hydrogen car will also only be a temporary solution until we have developed levitation. That's why we only discuss what we consider to be a long-term and sustainable solution for the future. But we can say this: to advance the hydrogen car beyond where it is today, we also need to introduce a magnetic propulsion capability. We need to continue improving some internal concepts that, in our view, don't fully work, including the gas itself.

As for the gas-powered car, it will be developed for use in specific applications, like on farms.

Archangel Metatron would now like to share, before we finish, about the new gas that you will create in the future. As usual, we don't provide a complete recipe, but there are plenty of clues scattered around.

HYDROGEN CAR

Archangel Metatron speaks:

The hydrogen car is currently a good solution, but eventually, you will discover a new path with a new type of gas that, together with magnetic propulsion, will power the car. This is a car that does not yet exist but will come into being in the near future.

This future car will generate electricity through a water-based component and create motion through magnetic capabilities, resulting in a gas that is completely harmless. However, this gas must be created in a manner similar to nuclear fusion, which we have already discussed. But it's not nuclear fusion itself that we will use; rather, it's the conceptual functions of nuclear fusion that will influence our future hydrogen car.

For us, nuclear fusion will be primarily used as an alternative energy source, while the hydrogen car will require a completely different technology because it will be powered by converted water vapor. Water won't be used in nuclear fusion, except for the occasional water vapor that might be produced. Therefore, a different technique will be employed, but we will still harness the capabilities of nuclear fusion in the future.

We know that today's hydrogen car already emits water vapor, which is why we advocate for the hydrogen car. However, it's the gas itself that we hope you will continue to develop.

ENTRY OF THE NEW GAS

God's spiritual guide speaks:

A new gas will emerge, one that "bubbles," and that's what we want you to explore now. How does this gas, this bubbling, come about? To explain, we need to return to the basics of protons, electrons, and neutron stars.

We've already discussed neutron stars and how we can use magnetism in our vehicles. But to achieve this movement, we need to revisit how we transform some fundamental concepts, such as electrons in an atom. According to our scriptures in the spirit world, an atom can only undergo a change if it's in motion. So, it's only when an atom is in motion that we can remove the electrons we want to remove. This technique can also be observed in black holes.

We've divided it into four parts:

1. Remove – transform electrons in an atom.
2. Fill the void that is created.
3. Supply.
4. Create a calm to prevent the gas from exploding.

When we remove electrons, a void is created. This void needs to be filled, and what we fill it with is called bubbles. Bubbles arise when we introduce new electrons, which will act more powerfully because these newly introduced electrons are much stronger than the old ones.

But to achieve these bubbles, we also need to add helium – you know, the gas we use in balloons. But in this case, we mix it with hydrogen atoms, from which we can remove electrons inside an atom and then create new ones. This is when a new gas that "bubbles" is formed. It appears to bubble, though it doesn't actually bubble in the conventional sense.

What else do we need to add? Well, hydrogen oxides. Oxidation is necessary here; otherwise, no bubbling will occur. Hydrogen oxide doesn't collide with other components because it acts independently, which is important since we don't want an explosion. But for hydrogen oxide to act with full strength, it needs something to enhance its capacity. This is achieved by adding magnesium oxides, including ions that are present on the moon. By adding magnesium ions (oxides), their power and strength increase.

Everything is drawn to magnesium's
ability to act with full force.
All can be utilized in the universe's
atmosphere as well.

We need to let everything oxidize, create ions, and then you will soon achieve an almost finished product that you can use in your machines. However, we also need protons. Protons will be addressed later, once we have researchers who can take on this new and improved technology.

Why do we need protons, and what good do they do on their own? Moreover, what do we do when the number of electrons increases? For a proton to grow, it needs a place to reside – an atom. But in this case, it doesn't get along with its newly introduced neighbor – the electron. That's why we need to separate them before they start to conflict.

In this situation, this is the case, but in other situations, you won't notice any difference between protons and electrons. But here, we will alter our internal structure because a problem has arisen, and by removing electrons, the atom's core will significantly calm down.

– Why do you want calmness? I ask God's spiritual guide.

– We need to create a peaceful internal environment before we can introduce something new, in this case, hydrogen oxide. You see, sometimes electrons and protons don't agree when we introduce new elements. There's a risk that they'll repel each other, disagree, and compete for the space they occupy. That's why we remove electrons, double them so they can expand and become stronger neighbors. After that, we can add hydrogen oxide. Oxides are important in this case; otherwise, no change occurs.

So before we create a new hydrogen car, we first need to ensure that we are using the right gas and that it's safe. The mixture might be a bit chaotic at first, but you will understand what we're advocating for when it comes to gas and cars. Gas can indeed be a powerful driving force if applied correctly and with the right ingredients.

There will be some puzzles initially, as hydrogen oxide may not be the easiest substance to develop an environmentally friendly gas from. But trust us, as you progress, you'll understand what we mean. It may sound crazy, but to us, it's the most obvious material to use in the future. That's why we need new scientists who dare to take on this new method with a new gas-driven medium. Some of them are already in place and are currently pushing this forward.

Ulrika: I will now write down the substances that I receive from the angels. Not in any particular order or with accompanying formulas, but just as they come to me, without fully understanding their significance.

Mg – An (or Ar) – O – sulfur (oxygen) – bubbles

Magnesium is used as an ignition component. Sulfur transforms into bubbles. These bubbles form a new gas that we can then use in our magnetically powered gas vehicle.

HNO3 – sulfur – hydrogen oxides – sodium chloride (ion)

Summary:

We remove electrons so that they can return, but in a new electronic form. Electrons are torn apart through a technique from neutron stars or even black holes. We then add new electrons and let them determine the speed. This is a self-generated process where the atom itself recognizes that a new protective shield needs to be created because the outside world poses a threat. These new electrons spin faster than the old ones to protect the core, which also increases the energy level at the center.

Protons knock out neutrons. The protons win but feel threatened, and when they do, new electrons are created as a defense – better and stronger. Think of how Mercury protects itself from the sun's intense rays with a strong internal and external shield. This transformation is necessary; otherwise, this new internal process won't start.

We also need hydrogen because it forms the base gas that the car itself will create. But to achieve this, we first need to harness the power of helium, which acts as a strong adversary. However, in this case, the adversary is a good leader, one that will guide the gas correctly.

So, we add helium to create the bubbles we talked about in the gas, through the oxygenation that occurs naturally. Helium increases, dilutes the oxygen level, and thereby also increases the gas we advocate for. Gas forms, like in the universe, spinning around, spinning together, which enhances its strength. But for all this to rotate, we need to tap into the core concept of neutron stars – magnetism. A so-called artificial neutron star has now been created with magnetic components at its center.

In the future, we will use magnesium more and more in our vehicles. And when it comes to our bubbles, a new technique will emerge, that of merging.

You might not understand this beautiful concept yet, but trust us, you will when the time is right and when your magnetic ability has developed as well. The magnetic ability is just as important as the gas we'll have then. So, try to achieve this new form of gas since it's non-explosive. And together with your magnetic movement, our world will be much better off.

So, the hydrogen car is a good model for now, but as we've said, it won't be further developed until you've mastered magnetism.

- Both the hydrogen car and the gas car will be considered outdated in the future. They will still exist to some extent, but more as a relic of a past era since some vehicles, like tractors or others, will still need to run on gas. But not regular passenger cars.

- We will also revisit old methods that some thought were just myths. But there were scientists back then who knew what they were doing, only to be overshadowed by others. Their work will be recognized again when we enter the era of magnetism. So, dig up those old scientists and see how they worked back then.

- The electric car will cease to exist entirely, and there are two reasons for this:

1. Mother Earth can neither act as an energy source forever nor should she. Eventually, you will deplete her strength, leaving nothing left to draw from. We want to avoid this.

2. We won't have electricity in the future because we will instead work with what the universe provides us with. But that said, we always need a *transition*, as we call it. It's difficult to evolve directly from A to Z. So, continue developing the electric car (without electricity) and the hydrogen car, as we see that they will eventually lead to the magnetic concept – the levitation car.

- By introducing magnetism, including levitation, we will achieve a cleaner environment, and our income will stretch much further since the power of the universe is much cheaper and simpler. Through magnetic movement, even your engines will become quieter.

- Electricity will be halted by environmental activists, especially when it is revealed that their electricity and purchases are not entirely clean. Their own sources will tire of secrecy and begin to speak out.

Today's electricity also poses a barrier to your future levitation.

KEFTU EXPLAINS

To make a car levitate, it's crucial that you create an entirely new form of energy, rather than relying on the electricity you have today. We have learned that your current electricity system isn't properly installed, as it uses a method that allows others to step in and disrupt a circuit or power supply at any time. This behavior is not in alignment with the laws of the universe, where no one should have the power to cut off a power supply that all citizens need. But don't worry, everything will eventually be set right again. As a result, they will no longer have the power to control it.

This is one of the things we, the Kathremes, have been working on for many years (on Earth) to help you transition to a cleaner energy system. This shift will lead to a new era where electricity is no longer the dominant and driving force. Very soon, as we mentioned earlier, other elements, including neutrinos, will emerge and fully take over that role. So, everything will change, both within and around you.

The electricity they've been using for many years hasn't been produced from a clean perspective, not as it's claimed to be. There's a lot more impurity in your electricity than they say, which hasn't come to light yet. They use a form of digitalization to manipulate and control this process, releasing particles that are harmful. This will also be discovered, but not until you start dismantling your power lines – or should we say, your entire electrical system. Then you'll see how dirty and dangerous your electricity really is.

We've also learned that the electricity in your electric cars isn't "directed" as it should be. By this, we mean that it doesn't discharge properly, which has been intentionally applied to your electric cars so that you have to recharge more often. That's why they've built in a stop in the car that doesn't show up on the display. But future researchers will find this stop through a diagnostic program, and then you'll understand what we mean – how it was integrated. They will then change the concept to a more accurate driving system.

So, you believe that your power supply is good, but it has only brought more pollution and destruction than you realize today. You'll discover this too, especially when you start lifting your cars. They have also blocked the power supply you need to be able to lift the cars of the future, which you don't fully understand yet. Here too, electricity has been used as a brake.

What we mean is that in order to levitate over a long distance, and not just go straight up, but also to drive the car as you do on the ground, you need something we call *free space*, where nothing *interferes* with the frequency. You will start a completely new power supply, a new energy frequency that you've never used before, and this will help you navigate. What happens when they lay out their power grids is that they block the universe's own power supply – or should we say, the universe's *free* power supply – and this is what you need when you start to levitate. It's what ensures everything runs smoothly.

So, it's the electricity that blocks this frequency, which is exactly what they want, because they don't want you to develop this levitation concept.

You've developed some trains, but have you ever wondered why they're still at ground level? The technology exists, yet you're not progressing. They know you'll never get a car to levitate as long as electricity blocks your exits because it's the electricity that powers the car. By "exits," we mean different stations that you need to anchor in the air because they are what allow the car to levitate over longer distances.

Think of it as different tunnels where each tunnel you pass through gives you more speed, which already exists to some extent on Earth today. But here we mean when this type of activity also takes place in the air. Because if you don't have this tunnel system, the car will remain stationary, even in the air. So, every time you pass through these exits, you enter a new time frequency. That's what it's all about when you start to levitate – different time cycles, different time frequencies that propel you forward, and so on.

This is what they want to prevent, which is also one of many reasons why they constantly want to expand their power grid.

Therefore, electricity has been intended to act as a brake, not to supply you with the means to live a good life as many believe.

In our view, electricity is also a very outdated technology. It's also much more dangerous than many understand today, especially if used incorrectly. And the more people wake up now and find their life purpose, the more people will reach this part of their development.

This is why it's so important that you uninstall EVERYTHING related to electricity before you begin developing a more advanced concept of movement for levitation.

Thank you, your dear friend, Keftu.

Kazandra speaks:

Now, my beloved friends, we won't go any further for this time. We, the Kathremes, thank you, and as mentioned, we will return when the time is right, and all contact will then be through Ulrika. Thank you as well, Ulrika, for the wonderful time we've had and continue to have with you.

We, the Kathremes: Alfredo, Kazandra, Bermuda, Afrodit, Fredzo, Freptzo, Captru, and Keftu. We all love being here, and we learn so much from you, and we long for the time when we can finally reunite with you.

Archangel Michael concludes:

Levitation will be the future concept, and the reason it hasn't yet taken hold is because some have put a stop to it. Therefore, we archangels will dissolve their negative activities before the year turns to three. Many of the participants will be revealed and exposed as barriers to the future.

It's also clear that some don't fully believe that we can introduce levitation. But this only creates a barrier for yourself and your development because we need to believe in a better future. As mentioned, levitation already exists, but before it can be fully released, we first need to hold some meetings, and that's where we will separate the wheat from the chaff. So, it will happen.

Therefore, we ask that you review your research stations, as your teammates need support in the work that is currently underway. And if you don't get the support you need, ask us archangels for help. We will then dissolve your stations and examine your leaders, especially if they don't provide you with the help and equipment you need to continue your work. So, you will receive help and support in this important work that lies ahead.

Thank you from us, your angels and the beloved heavens, and Ulrika, our pen-writing soul.

APPENDIX

WHO ARE WE?

Ulrika begins:

It's a spring-summer day in 2020, and I'm sitting and enjoying my garden when I suddenly receive new signals within me. At first, I thought it was my tinnitus, so I leaned my head against the back of the chair and continued to relax. But the signals persisted and grew more intense, and since I hadn't heard them before, I became worried that my tinnitus had worsened. So, I reached out to my guardian angel, Nikodemus, and we began writing together. He explained to me where the signals were coming from.

– They are signals from other souls who call themselves Kathremes. It is written in our soul contract that at this time, we will collaborate with the Kathremes and help them convey the information they have come to share. They are friends we will work with in the future, so welcome them into your network, my father said.

– I will, Father. But who are they? I asked.

– We'll let them introduce themselves. They will contact you when you meditate.

And that's exactly what happened. I started writing with a Kathreme named Bermuda. She also told me that I had lived with them in one of my past life's, which we will discuss later. As time went on, more Kathremes came forward. Now, we want to share the conversation when the Kathremes introduced themselves to me.

I'm sitting at my kitchen table, and Kazandra enters:

Hello Ulrika

We are the Kathremes, and we are all so happy to be here to assist you in the work and changes that need to happen now before you can fully step into the new Earth. It will be a place where we can finally live together as friends, something we have all longed for so much.

Who are we? As mentioned, we are the Kathremes, and those who will step forward in this are Alfredo (captain), Kazandra (myself), Bermuda, Afrodit, Fredzo, Freptzo, Captru, and Keftu.

Alfredo

Alfredo is our captain. He has been with us for many years and has taken on the commitment to fight for peace so that everyone can reach their full potential. Because of this, he travels frequently in his work, which sometimes makes it difficult for him when he has to leave his wife and two children.

He might seem a bit rigid at times and strict in his ways, but it's only because he genuinely cares about everyone and wants to ensure that no one is exploited. If you oppose him, you might find yourself reflecting in a rather stern face. But he guides with discipline. He leads based on what he knows works and never takes risks to discover new things. That's why we always conduct many tests before sharing a new discovery.

His goal is always unity – ensuring that everyone can share in and work toward the same objective. That's why we have revived our connection with Ulrika, our soul mate, because it is an old connection we had from before, where she lived with us as a student to learn the foundational knowledge we had at that time.

We also know that Alfredo has already accepted Ulrika as a new, loyal family member. He says that she works well. She wants to develop and learn, even though her body sometimes protests. But as long as the desire and will to work for peace and love are present, you will always have our captain by your side as a faithful companion.

We are all so proud to have him as our captain.

Kazandra

Regarding myself, I still hold onto my teachings, even though I have learned for many years. But if you want to develop and become fully realized, your teachings must never become stagnant. We continuously take in new discoveries and always learn from the old – revisiting and learning anew, which we also wish for you humans to do now. You have become locked into your teachings, which you will soon discover.

I am a righteous woman who has lived many years. I am soon to be what you would call a retiree. Yet, I am still on this journey to contribute the knowledge I have gained over the years. Since I am about to conclude my work, I took on this mission as a wonderful opportunity to make one final journey with my companions and to meet Ulrika, our new friend.

Thank you, Ulrika, for everything you do. We are all already so fond of you. Your soul shines in the color of gold when we write together, which for us signifies great loyalty. Thank you, my dear friend.

Bermuda

Bermuda is the one you would call a nurse, but in our country, we always focus on the psychological aspect first before moving on to the physical. This is because we can avoid many physical injuries if we first learn to manage the inner self properly. Therefore, you too have something new to learn here. It's important to address the psychological aspect more thoroughly than you currently do, especially those of you working within the healthcare field. Bermuda will assist you with this as we embark on our journey together in the future.

Bermuda can also provide recipes for healthful beverages, which are among her specialties on the mothership. These are drinks that give you strength and heal your body without adding artificial substances. We are happy to share these with you, if you are interested. We will be staying here for a while to help you in any way we can. But when the time is right, some of us, including myself, Kazandra, will return home, while others will remain. We will never leave you!

Afrodit

My name is Afrodit, and I am a guide on the mothership. It is my role to ensure that everyone follows the plan we have set for this journey.

We have become aware that many people do not listen inwardly, and this is what I am here to help you with – so that you can reconnect with your inner self, stand strong in your own power, and stop heeding empty words from your leaders. I have taken on this significant task to reach out to you and understand what you need to do to start working together as a united people, rather than fighting and going against each other.

There will also be a liberation from those who have long kept you captive, and this is where my work comes in. I also have a military background, but not in the way you might think. In our current context, we guide people into what is your life. We no longer have any war preparations, only some outdated ones that remain as reminders of a time we should not return to. Therefore, we always work based on each individual's needs so that you can develop within your discipline.

It is my responsibility now to lead you toward a united society where you work together. We start our work with those who resist before we can move in and make changes. This is also something we have already begun.

I now hand over the word to our mechanical workshop. Thanks for me, Afrodit.

Fredzo

My name is Fredzo, and I am the chief mechanic on the mothership. I want to share with you the challenges you will face before the year turns to 2025. If you are not awakened when the great change occurs, you may become confused, which is common, so don't worry. This usually happens when we clear out what was previously full – something you have believed in for a long time, which then turns out to be false. As a result, some may feel deceived. They will then continue to try to mislead you. Do not listen to them! Go against whatever they say! Do not heed their words; this is our advice on the matter.

We are here to assist you with this, and we will come down and stand by your side, but not until the time is right. Therefore, it is crucial that you remain calm. For this to happen, we need support, not chaos. You will then see how things have really been with you. As a result, some will act out their despair, revealing their existence and what they have done. When this happens, it is important that you do not erupt into a storm but support each other, as that is what makes you strong.

Much needs to happen, even within you, before we can land among you. Among other things, you need to shift your color, as we like to say for a bit of fun. What we mean is a higher frequency so that you better understand the journey we will make together in the future.

So be prepared, as we will gradually reveal ourselves on a larger scale, and many of us now stand united by your side. More ships have arrived in recent days (this was in 2020). Therefore, our immense strength might scare some at first, but only initially. Once this is revealed, some will attempt to question our credibility and the teachings we are here to convey, but only for a short time. So if you see a blinking light in the sky, it is us sending a greeting, so you know that we are here and supporting you in all that you do. Do not be afraid. We have come in peace.

Freptzo

Hello, my name is Freptzo, and I work on one of our nearby ships, not far from the mothership. I am what you might call a second-class sergeant. Alfredo, the captain of our mothership, has now given us permission to share our location with you.

My friends and I, those of us on the smaller ships, are 600 miles from Earth. So you see, it's not as far as you might think. We have also been here for many years to observe and learn as much as we can so that we can assist you in the best possible way. As Fredzo mentioned, we will soon be entering your world, but we cannot reveal how this will happen. However, within a few years, more people will meet us even in our physical forms.

So do not worry if any of our friends attempt to contact you. If you feel uncertain, ask Ulrika on our shared channel, "The Kathremes." She will help and guide you. Many will also feel that we are not as foreign as some might think because, like Ulrika, you too have been in contact with us in another life. Perhaps you might even recognize our energies.

Feel free to call on us by our names if that is significant to you. Otherwise, Kathremes will work just fine, and the one responsible for your case will come forward.

Captro

Captro is a very distinguished man. He, along with Keftu, will be the ones to lead us down to Earth. His role will be to unite us with you and ensure everything goes smoothly. He was also the first to contact Ulrika in a more physical sense, including when we started sending a few messages on our YouTube channel about a year ago. However, we then took a long break because the timing wasn't right. Captro will also be the one to guide us into a new era.

Keftu

Keftu is the son of Bermuda, and he is a proud son. Despite his young age, he is already a fully-fledged leader, and he has also been in contact with Ulrika via his telepathy. He was the one who stood by her side and tried to teach her about our way of conveying information. Even though it didn't fully succeed at the time, we know she will master this type of communication excellently in the future. So it will happen.

Kazandra continues:

As you might understand, we all have different qualities, but we are also very similar, regardless of gender, so don't confuse our names with anything else. This is something we will discuss later: that we are a unified people and that everyone is equally important. It is humanity on Earth that weakens humanity. For you all come from the same part, but then you create divisions. This is something you need to strengthen and start seeing all individuals as equals – women and men alike.

– Where do you come from? I ask Kazandra.

– We come from a place not far from Orion, where the star Sirius shines so brightly. It is also where we learn and develop magnetism. Our planet is called Thrinica 2. We have two planets: Thrinica 1 and Thrinica 2.

– Why do you have two planets?

– Thrinica 1 is a base planet for our machinery. It is a base where we operate our machines. But it wasn't always like that. We actually discovered Thrinica 2 when we began flying. It was a larger planet, and as our population grew, we needed a new place to settle. So we began a new journey on Thrinica 2. That was where our ancestors started their journey and built the world we have today.

Since our world developed, so has our technology, and that's when we developed the magnetic concept. Therefore, we now have knowledge that only a few galaxies have today. It is also important that we share our knowledge so that we can finally begin to live and grow together, and that you can also become part of the world of magnetism. You will no longer use electricity in the future.

The technology we use on our planet is called *The Magnetic Sphere*. It is a fundamental technology that you will also use in the future. We have explained this in our book. But the technology we use works for us because both our atmosphere and its associated magnetic field are slightly different from yours. Our planet actually has two magnetic fields. The core of our planet is so strong that it can support this.

We have a large magnetic field that surrounds our planet and is in direct contact with us. This is what we use to interact with each other. We also have an external overlaying magnetic field, which acts as a shield against the strong gravitational fields that sometimes affect our world.

Therefore, we work where you are, not where we are, so you can act based on the power you have on your planet. This is written in our soul contract, the one we have with your God. Therefore, we cannot override you, and all development must take its time. But we are allowed to share some news, though only in small amounts and only if your God has given permission for it. These are teachings we once had to learn when our own nest was nearly depleted. It was a time when our own seat of power faced a very difficult task, just as you do now. Therefore, we will now stay with you for a while to help you in the process that needs to happen.

– What language do you use?

– We use "The Kathremian Language." It is an allied language that many already use in the universe.

– Where does it come from?

– It comes from Kathedrania, the place where we operate. We call ourselves "Earthbound Kathremes." We usually say, "I am a Kathreme".

– Are there non-Earthbound Kathremes?

– There are, but not in the way it sounds. It is another people who are not in the same place as we are. But to differentiate ourselves, we call ourselves "The Earthbound People of Kathedrania."

– I have learned that I lived with you in a past life. Where did we live then?

– We lived on a place called Dionsys, which means connection. It was a place where others could stop over. It was also where we manufactured and repaired various contemporary machines, including magnetism.

– How can we communicate with each other when you are there and I am here?

– It happens through telepathy. But it also occurs through the spirit world. It's like a halfway stop where your angels take over and translate our words. And since your telepathic ability is so highly developed now, it is what enables us to communicate with each other. Bermuda would be happy to explain how you learned telepathy from us.

– Oh, thank you, I would love to hear that.

Bermuda speaks:

In order for us to connect with each other in this lifetime, it was important that we had a method of communication, in this case, telepathy. Ulrika also learned about this early in her career, and it is what we have further developed now.

– How did I develop my telepathic ability with you? I ask Bermuda.

– Before you came to us, you had already acquired an inner teaching through past life's and the experiences you brought with you from them. For instance, when you lived as a powerful shaman woman who was not at all afraid to assert herself and use her inner power. This is also what we need to rekindle in this life, both the teachings you learned then and the inner power we know you have.

After that, you undertook several personal journeys where you lived some sorrowful lives. Just like in this one, where you once again faced your inner self and learned to handle everything that happens on the outside. But we won't delve into this now; it has been a long journey to get to where we are today.

Eventually, you landed with us. There, you not only worked with the magnetism we had at the time but also complemented your teachings regarding your spiritual concept, and this is where telepathy comes in. Your telepathy could not develop until you had completed your inner journey. That's why it is so important for everyone to understand that we cannot develop solely through tools, as spiritual development always happens through what you do on the inside.

– How did I come into contact with telepathy?

– We placed you in a school where you learned some sacred texts. These are also texts we will bring to life later on, perhaps in the next book. But it all started with us building up your sensitivity. How you were trained from childhood to be attuned to your surroundings. You had already developed a keen eye at that time. This was what was awakened in you early in this life, so it could develop at this time. That's why in 2016, when you began writing with your father, you could write and write without knowing it was actually automatic writing happening then.

As your sensitivity increased, we had you start writing earlier than others. You had to feel what you wanted to write about and try to follow that thread. This is how you learned to embrace your higher self.

Once you mastered the art of both inward listening and reaching your higher self, you examined other texts. These were texts you had access to early on and that are available on our planet. They were texts where we wrote about the magnetic ability, and you learned to connect what was what. That is, to use your sensitivity and learn where the text came from. Was it from your higher self, from another highly developed soul, or from another dimension?

You often stayed up at night to embrace your teachings because, as you said, it was quiet then, allowing you to feel and hear your and others' sensitivity. This is also why you pick up others' sensitivity and their emotions, which then disrupts your inner channel. We will help you with this so you can remain and embrace the silence you need to work and feel good.

That's how it all started, and before you turned 30, you had developed into a fully-fledged Telepathist, as we call it. Therefore, we are now here to rekindle the friendship our people had with you at that time. And the fact that we can communicate with each other today is through the technique you learned from us.

But physically, to seal our contact, we first needed to create an inner love, an inner bond. This is so that we always first unite in our hearts before we can do so in our thoughts. And we know that you from the beginning felt our authenticity and trusted that we are who we are, which is truly fantastic. So thank you for your trust. We know it can be difficult for you to take in the right information from us Galactic's and trust that it is true. But you sensed early on that it was so, which is why you never hesitate.

Our communication is based on telepathy, which is a high-energy frequency of thought processes and always starts from within, never from outside. As mentioned, it is in your heart where the feeling for the words exists, the ones you hear. Sometimes we have direct contact with Ulrika. She picks up the feeling in her heart and senses that something is happening. Then come the words.

But sometimes we also need to store information, which she then retrieves when she can. It's not always possible for us to write when we meet. Instead, we can place our message in a kind of container at that energy frequency, which she then opens with her permission, and the text we placed there comes out. Think of it like a recording. You record a movie and store it on your TV to watch the next day.

This is a highly advanced technique that we taught Ulrika because it is important that we can do this when we provide her with text she cannot receive at the moment. When she then opens the recording, we receive a message that the storage has been emptied.

We can then add new text that she opens and writes down. This continues. Sometimes we are in the present, but sometimes we are at an energy frequency where we store information.

– How can we open a stored text?

– To open the message we have placed, you need to reach the same frequency. That's why we sometimes need to raise your energy frequency, and this is what your inner journey has created. The journey you undertook before our first contact was established. When you reach that, you retrieve it in a private space. It's a space that is intended solely for us.

– Where is this space?

– It is in contact with the energy level where your soul resides. So you see, the more you develop, the more and higher you will connect with.

– So we can retrieve information like from a private vault?

– Yes, exactly, and it is at the level where you are.

– Do you have to lower your level then?

– Yes, but not necessarily, because we are not talking about different dimensions. We mean the energy frequency through which we communicate and the link we use. We can store information in that link and collect everything like in a vault. So yes, we can lower ourselves but do not always need to.

It works a bit like this for you when information is stored in your higher self. You can then retrieve the information stored there when you are ready to receive it. But here it happens within the same frequency, not in a higher near memory. We always communicate through energy frequencies when using a high telepathic ability.

– Thank you, my dear friend, now I understand.

– You're welcome, beloved friend. We will further develop our telepathy and the life you lived then in our future books. We also have a Kashan with us, and he will help you in the future, Ulrika. A Kashan is a teacher in the Kathremian language. He is a very good friend to us, those of us who come forward in this. And he has plenty of patience, he says, so don't worry (smiling and winking). (I laugh heartily because the Kathremes know by now that I have very little patience and would prefer everything to happen yesterday.)

– Thank you, I understand and I look forward to it.

– Thank you, so do we.

We also have plenty of shamans on our planet. These shamans teach about all the arts and rules of the universe, which you will also do in the future. They will bring forth new knowledge, and this is what you will need to learn when the year of the shaman comes to you. They will attain a high status on your planet, not to consider themselves superior but to assist you in the process that will occur then.

They become like doctors when they help a person in need, but in this case, they assist your planet through the power provided to us by the universe. Therefore, there are many shamans on Earth today who are already speaking about this and have established good contact with both the spirit world and other beings found throughout the universe.

As we mentioned earlier, Ulrika has studied the path of the shaman during her time as a shaman woman in the USA in one of her past lives. Therefore, she has now been elevated to the source of the universe to explain and help you with what you need assistance with.

It was significant that we received permission from your God to use some of your beautiful Earth angels, light workers, who are here to work for the good light. Ulrika is an Earth angel who has helped us before, so we welcome her back into our world. That's why our contact has been so smooth. And the lives we lived together then will, as mentioned, be further developed in our future books.

We Kathremes are a proud people, and we already see you as part of us.

We love to sail against the wind, and we always
work to create a good balance in the universe.
We are the universe. So are you. We are all one!